Endorsements for *The Renewable Revolution* by Sajed Kamal

'Sajed Kamal's book is an immensely important contribution to our understanding of what is happening to our planet and what we can do about it. He combines the careful research of the scientist with the passionate imagination of the poet. I hope *The Renewable Revolution* will be widely read.'

Howard Zinn
Author of *A People's History of the United States* and
You Can't Be Neutral on a Moving Train

'In this book, Sajed Kamal reminds us what we intuitively know – that we must make the transition toward renewable energy as soon as is humanly possible, that continuing to rely on fossil fuel poisons our world with every hour that passes. Even more powerfully, though, he reminds us how possible that transition is. Equally competent with a spreadsheet and a bank of batteries, Kamal is the navigator we need to sail with confidence into this new century.'

Bill McKibben
Founder, 350.org. Author of *The End of Nature, The Age of Missing Information* and
Fight Global Warming Now: The Handbook for Taking Action in Your Community

'Dr Kamal lives and writes a vision. It is a vision of reconciliation between humanity and the planet – a reconciliation that is accomplished through the medium of energy – natural energy, spiritual energy, clean energy. In plumbing the implications of solar technology for our survival, Dr Kamal has identified the transcendent nature of that technology. Solar energy, he reveals, is far more than a simple source of electricity and warmth. It provides an elemental bridge between a history which has been based on depletion and exploitation and a future which, if it is to exist at all, must be based on balance, renewal and a most practical harmony with our species' home.'

Ross Gelbspan
Pulitzer journalist. Author of *The Heat Is On: The Climate Crisis, The Cover-Up, The Prescription* and *Boiling Point: How Politicians, Big Oil and Coal, Journalists and Activists are Fueling the Climate Crisis – And What We Can Do to Avert Disaster*

'The great value in this book is the picture it conveys of renewable energy today: the long sought transition is no longer a hope for the future, it is already underway. This is a fine survey of the many applications of solar technologies.'

Howard Ris
President, Union of Concerned Scientists (1981–2003)

'An important tale that has to be told and Dr Kamal tells it with the passion and precision that's needed. But *The Renewable Revolution* is more than a message – solutions are offered and it's our job to follow the solar course so elegantly set for us.'

Jane Weissman
Executive Director, Interstate Renewable Energy Council. Former Chairperson,
Policy Committee and Board Member, American Solar Energy Society

'*The Renewable Revolution* is an excellent treatise on the practical use of renewable energy. Sajed Kamal has not only pleaded for the worldwide use of renewable energy, but also proven its practical utility in supplying energy to promote sustainable development of communities. He has proven with illustrations, many from his own experiences, that harnessing renewable energy is not only environmentally sound, but also economically viable and socially acceptable.'

Kazi F. Jalal, PhD (Harvard)
Faculty, Harvard University Extension School Environmental Studies Program. Former Chief, United Nations-Economic and Social Commission for Asia and Pacific (UN-ESCAP). Former Chief, Office of Environment and Social Development of the Asian Development Bank (ADB). Co-author of *Introduction to Sustainable Development*

'In the face of the growing reality of climate change and its devastating worldwide consequences we need solutions. An urgent transition to the renewable energy path is fundamental in that solutions mix. *The Renewable Revolution*, an exceptional combination of scholarship, personal experience and inspiring examples, charts the path. This comprehensive and highly readable book is both an indispensable educational resource and a tool for action.'

Saleemul Huq, PhD
Director, Climate Change Programme, International Institute of Environment and Development (IIED), London. Lead author of International Panel on Climate Change Report, recipient of the Nobel Prize for Peace, 2007. Recipient of the Burtoni Award for contributions to Research on Climate Change Adaptation, 2007

'Dr Sajed Kamal proves once again that he is a most thoughtful and knowledgeable advocate of sustainable solar energy and its critical potential contribution for our future. *The Renewable Revolution* needs to be read by decision-makers at the World Bank, the UNDP, as well as the smallest community-based organizations if the costly mistakes of current policies are not to continue to pollute and impoverish us, threatening our sustainability.'

Prof. Laurence R. Simon
Professor and Director, Program in Sustainable International Development, Brandeis University, Massachusetts, US. Former advisor, the World Bank and the UNDP

'At this time of darkness when most people have given in to despair, apathy and powerlessness, Sajed Kamal's book is a true beacon of light. It is hopeful, inspiring and practical. It shows the way to the resolution of the global energy crisis as well as the deeper moral crisis in which the so-called environmental crisis is embedded. A powerful and compassionate work.'

Asoka Bandarage, PhD
Visiting Professor, Elliot School of International Affairs, Georgetown University. Author of *Women, Population, and Global Crisis: A Political Economic Analysis*

'Clear, insightful and compelling. *The Renewable Revolution* provides us with inspiring examples and information to understand how renewable energy is succeeding, together with the moral vision to realize that we must.'

Bruce Allen
Professor, Department of English Language and Literature, Seisen University, Tokyo, Japan. Environmental writer and translator of *Lake of Heaven* by Ishimure Michiko

'All life is one – interconnected and interdependent with the rest of Nature. We can survive only if this right understanding is acquired and cultivated. Sajed Kamal conveys this message where knowledge, wisdom, intuition, experience and vision are all combined with a clear agenda for the novice as well as the expert. This one book contains everything that one may seek in one hundred books.'

Dr A. T. Ariyaratne
President, Sarvodaya Shramadana Sangamaya, Sri Lanka. Recipient of the Magsasay Award, The Philippines; The King Baudouin Award for International Development, Belgium; Niwano Peace Prize, Japan; Hubert H. Humphrey Award, US; Gandhi Peace Prize, India

'*The Renewable Revolution* is not just another book on alternative energies: the author is a scientist and a poet as well. He not only started 'solar revolution' projects in different parts of the world, he also lives what he preaches. He brings 'Head and heart and hands together,' as the Swiss pioneer in education, Heinrich Pestalozzi, suggested. All this shines through in this very well written and richly illustrated book and makes it an outstanding tool in this crucial time of change.'

Martin Vosseler, M.D.
Founder of PSR/IPPNW Switzerland (Swiss chapter of the International Physicians for the Prevention of Nuclear War, Nobel Peace Prize, 1985). Co-founder of the Swiss Physicians for the Environment and of 'sun21'. Crew member of Transatlantic21 on the first Atlantic crossing with the solar catamaran 'sun21' (*Guinness Book of World Records*, 2007). SunWalk 2008 from Los Angeles to Boston. Recipient, European Solar Prize 2007

The Renewable Revolution

The Renewable Revolution

How We Can Fight Climate Change,
Prevent Energy Wars, Revitalize the Economy
and Transition to a Sustainable Future

Sajed Kamal

publishing for a sustainable future

London • Washington, DC

First published in 2011 by Earthscan

Earthscan Ltd, Dunstan House, 14a St Cross Street, London EC1N 8XA, UK
Earthscan LLC,1616 P Street, NW, Washington, DC 20036, USA
Earthscan publishes in association with the International Institute for Environment and Development

For more information on Earthscan publications, see www.earthscan.co.uk or write to earthinfo@earthscan.co.uk

ISBN: 978-1-84971-195-1

Typeset by Safehouse Creative
Cover design by Rob Watts

A catalogue record for this book is available from the British Library

Library of Congress Cataloging-in-Publication Data

Kamal, Sajed, 1945-
  The renewable revolution : how we can fight climate change, prevent energy wars, revitalize the economy, and transition to a sustainable future / Sajed Kamal.
     p. cm.
  Includes bibliographical references and index.
  ISBN 978-1-84971-195-1 (hardback)
  1. Renewable energy sources. 2. Sustainable development. I. Title.
  TJ808.K36 2010
  333.79'4--dc22

                            2010008343

At Earthscan we strive to minimize our environmental impacts and carbon footprint through reducing waste, recycling and offsetting our CO_2 emissions, including those created through publication of this book. For more details of our environmental policy, see www.earthscan.co.uk.

Printed and bound in the UK by T J International, an ISO 14001 accredited company. The paper used is FSC certified and the inks are vegetable based.

Mixed Sources
Product group from well-managed forests and other controlled sources
www.fsc.org Cert no. SGS-COC-2482
© 1996 Forest Stewardship Council

Contents

Synopsis

Our world faces an unprecedented energy crisis. Fuel shortages, skyrocketing energy prices, climate change, nuclear contamination, catastrophic oil spills and energy wars define the global scenario. The nonrenewable energy path of oil, natural gas, coal and nuclear is headed for a dead-end at an accelerated speed. In our race for survival we are awakened to the simple truth that the essential condition of sustainability lies in our ability to live within the limits and renewability of natural resources. It invokes within us an urgent need for transition – from an obsolete, destructive and unsustainable energy path to a sustainable path of innovation, renewable energy and peace. The good news is that the technology required to make this transition is already available.

From an author with over 30 years of experience campaigning for and setting up renewable energy projects around the world, this book is unique for its interdisciplinary approach – interweaving technology, economics, environmental science, philosophy, history, spirituality and politics, asserting that to understand the crisis and find a sustainable solution requires a holistic perspective. Readers will understand the vast renewable resources we have at our disposal in the form of sunlight, wind, heat, water movement and photosynthesis, and the technologies used to harness this power. There are also the emerging prospects of solar hydrogen fuel cells, biofuels and geothermal systems. The true economic advantages of a shift to a renewables-based economy (and how we can get there) are laid out clearly. There's much to learn from examples around the world while we devise local and appropriate solutions.

Written for a crossover readership of students, educators, professionals, academics, activists and policymakers, both nationally and internationally, this is a comprehensive but readable and practical book that will inspire people to wake up to renewable solutions.

About the Author

Sajed Kamal, EdD, who teaches 'Renewable Energy and Sustainable Development' in the Sustainable International Development programme at Brandeis University, has been involved in the field for more than 30 years. He has been a lecturer and consultant on renewable energy internationally, setting up projects in the United States, Bangladesh, Sri Lanka, Armenia and El Salvador. His work has also provided the basis for projects in Latin America, Europe and Africa. He is also an award-winning poet, artist, educational consultant, psychotherapist, translator and published author of a dozen books and many articles in a wide range of areas. In 2007, he was awarded the Boston Mayor's First Annual Green Award for Community Leadership in Energy and Climate Protection and, in 2008, a Lifetime Achievement Award by the U.S. Environmental Protection Agency, New England Region. The 'Greener Issue' of *The Boston Sunday Globe Magazine* on 28 September 2008, featured him as one of the 'Six local heroes whose work is having rippling effects – at home and far away – in making the world a better place'.

Acknowledgements

Growing out of more than 30 years of experience in the renewable energy field, this book is a product for which I am deeply indebted to many persons and organizations around the world that have inspired, challenged and supported me in my writing it. The list would be too long, so I will only be able to mention a few. Even in doing that there may be some inadvertent omissions. I will include them in any future editions.

A vision, an idea or a hope for a solution is not much more than something in one's mind – unless it can be implemented and put to the test. That involves others – their interest, opinion, participation, collaboration and partnership. I thank all my colleagues – in the US, Bangladesh, Sri Lanka, Armenia and El Salvador – for being there for me. Nothing has taught me more – practically, intellectually, socially, politically, morally – than working first hand on actual projects involving the issues and technologies I write about.

I thank all my colleagues and students at the universities where I have taught – Boston University, Northeastern University, Antioch New England Graduate School and Brandeis University. While teaching, I have also learned – that's the essence of dialogue. The same is true with my lectures before highly diverse audiences.

I thank the journals, magazines and newspapers that have published my articles on the subject, paving a path for this book. Among these are *The Fenway News, New England Condominium,* the on-line journal of the Interstate Renewable Energy Council (IREC USA), the *Journal of Bangladesh Studies* (a peer-reviewed journal published by the Bangladesh Development Initiative, www.bdiusa.org), *The Daily Star* (Bangladesh) and *New Age* (Bangladesh). In Bangladesh, under the leadership of Farhad Mazhar and Farida Akhter, UBINIG: Policy Research for Development Alternative published my booklet *Photovoltaics: A Global Revolution & Its Scope for Bangladesh* (1989). It proved to be highly instrumental in my simultaneous launching – and subsequent promotion – of 'A Photovoltaic Pilot Programme for Bangladesh,' with the implementation of 10 stand-alone PV systems across the country, pioneering the utilization of the technology in the country. I thank the Overbrook Foundation for a grant for the programme. Some of the Boston projects which I have initiated, and which I describe in this book, were funded by the Mission

Hill Fenway Neighborhood Trust, Massachusetts Technology Collaborative, Solar Boston, Fenway Community Development Corporation, Massachusetts Energy Consumers Alliance, City of Boston and Boston Public Schools. I thank them too.

As the manuscript evolved, I invited comments. Many people offered their comments and I am grateful to them all. They include the endorsers. Not only did they enthusiastically take the time to read and comment on it, some of them offered valuable critiques toward finalizing the draft, as well. They are: Howard Zinn, Bill McKibben, Ross Gelbspan, Howard Ris, Jane Weissman, Kazi F. Jalal, Saleemul Huq, Laurence R. Simon, Asoka Bandarage, Bruce Allen, A.T. Ariyaratne and Martin Vosseler. Here I would like to say a few words about Howard Zinn, whom I first met as a doctoral student at Boston University in the early 1970s, and later over the years at numerous anti-nuclear and anti-war rallies. His sudden death on 27 January 2010 is a loss to the world never to be recovered. Rosie, my wife, and I saw him coincidentally on the street last year and I was happy to tell him that the manuscript was under consideration by Earthscan. He was delighted and had only encouraging words to say. It saddens me deeply that I won't be able to present this book to him. But – while feeling deeply honoured to have his endorsement – I hope this book will also serve to carry his voice of conscience, justice and peace to its readers.

I thank the photographers. I have credited each photo individually, either by the photographers' names or by mentioning the organizations through which I got the permission to use the photographs. If I have inadvertently left out anyone, I apologize and will make corrections for any future editions.

The awards I have received have certainly added credibility to my writing. For honouring me with such awards I would like to thank Boston's Mayor Thomas Menino, Chief of the City's Office of Environmental and Energy Services Jim Hunt, Director of Energy Policy Brad Swing, Special Assistant Mayor's Office Sarah Zaphiris and City Councillor Mike Ross; the U.S. Environmental Protection Agency, New England Region; Fenway Community Development Corporation; AltWheels: Creating a Sustainable Transportation and Energy Vision for the 21st Century; and the New England Bangladeshi American Foundation. Also, there are numerous magazines, both in the US and abroad, who profiled me and television and radio stations who interviewed me. I thank them all.

I must also thank the following individuals: Dr. Sultana Zaman, Professor Muhtasham Hussain, Khushi Kabir, Rotarian Anisul Islam, Dr. Muhammad Ibrahim, Rayeesuddin Bhuiyan, T. A. K. Naseem, Obaidul Quader, BRAC Chairperson Fazle Hasan Abed, Deputy Director Aminul Islam, Sharif Uddin, Sudip Saha, Mahbub Alam, Jainul Abedin, C. N. Quader, Afroz Rahim, Niaz Rahim, Dr. M. Eusuf, Dr. Atiq Rahman, Dr. Shahidul Alam (Zahed), Prof. Saiful Huque, Prof. Zahid Hasan Mahmood, Sultana Alam, Prof. Anu Muhammad, Asadullah Al-Husain (Humayun), Saiful Huq Omi, Hayat Imam, Badiuzzaman Nasim, Mridul Chowdhury, Mushtahid Ahmed, Quazi Nuruzzaman, Akram Bhuiya, Amjad Hossain, Nancy Anderson, Gurinder Kaur, Susan Kinne, Steven Strong, Osla and Bob Case, Lena and Richard French, Carolyn Mugar, John O'Connor, Regina Eddy, Anna Edey, Michael Kalashian, Arthur Hairumian, Steve Minkin, Harry Goldman, Mary Canning Goldman, Sean Ahearn, Avi Davis, Joan Schwartz, Gloria Carrigg, Richard Hansen, Marianne Wolpert, P. M. John, Ifeanyi and Carol Menkiti,

Kathleen Spivack, Malcolm Brown, Andrew Stern, Larry Chretien, Martin Schotz, Neela de Zoysa, Steve Fernandez, John Payne, Anis Ahmed, Syed Ziaur Rahman, Charl Maynard, Ellen Browne, Rumi Shammin, Marianne Buscemi, the Vandermark family – Henry, Liz and Juliana, Mary Essary, Ambrose Spencer, George Mokray, John Francis, Tim Harkness, Warren Leon, Dr. Roger Dickinson-Brown, Janaki Blum and her family – John and Rohan, John Lauerman, the Noce family – Rich, Margaret, Greg and Nate, Shawn Reeves, Julie O'Neil, Werner Grundl, Jim Lance, Erica Flock, Dilara Afroz Khan, M E Chowdhury Shameem, Amal Akash, Nahid Nazrul, Zafar Sobhan, Carl Nagy-Koechlin, Shelley Dein, Lisa Soli, Jack Neuwirth, Jethro Heiko, Steve Brophy, Elaine Yoneoka, Mitsue Allen, Alex Campbell, Sue Shenkman, Marc Gurvitch and Beth Mahar. Thanks also to all my fellow committee members and associates not mentioned thus far at BASEA (Boston Area Solar Energy Association); International Consortium for Energy Development; Solar Fenway – Samina Ali, Peter Flannery, Kate Janisch, Ted Joyce, Punita Koustubhan, Marc Laderman, Kathy McBride, Mandy O'Brien, Richard Orareo and John Pearson; the Fenway Community Development Corporation; Democracy and Development in Bangladesh Forum; Bangladesh Development Initiative; BANE (Bangladesh Association of New England), BEN (Bangladesh Environment Network), and BAPA (Bangladesh Poribesh Andolon/Bangladesh Environment Movement).

It is my absolute good fortune to be published by Earthscan – to be joining in our shared mission and vision for a healthier, just, peaceable and sustainable future. It has been both a pleasure and privilege to work with Michael Fell, my editor. The book simply wouldn't be what it is without his commitment, knowledge, astute observation and guidance. Working with him has been a mutually respectful dialogue that has both supported and challenged me, leading to clarity, assertions and changes, when appropriate. Not only have I felt that the integrity of my voice and writing was fully valued and retained, it was also enhanced. The rest of the Earthscan team that I have been in contact with – Nicki Dennis, Anna Rice, Martha Haworth-Booth, Claire Lamont, Lee Rourke and Martha Hawley-Bertsch – have demonstrated to me exceptional professionalism and helpfulness. Rob Watts has designed an attractive cover and Howard Watson has done skillful copy-editing. I am grateful to them all.

My deeply cherished friendship with Alex Roth – spanning over 47 years since our Loomis prep school days – is a testament to the saying: 'A friend in need is a friend indeed.' It's not the first time that Alex has been helpful to me in preparing my manuscripts. And I am sure, this won't be the last. Use of the computer is something I try to keep up with; Alex masters it. Especially for the inclusion of a whole chapter of photographs – and preparing them according to the publishing criteria and transferring them overseas, this manuscript presented a special technical challenge for me. Another challenge was to find someone with the skill, meticulousness and an understanding of the subject to prepare the index. Unfailingly – and ever cheerfully – Alex was always there to help. Even more than that, his comments rooted in our shared concerns, his intellectual depth and knowledge, and his eye for detail enriched this book immensely. Thank you, Alex!

Last but not least, my loving thanks to my family members both in the US – Rosie and our son, Ashok – and in Bangladesh – my brother, Shahed Kamal, brother-in-law, Supriyo Chakravarty, sisters, Amena Quahhar, Sultana Kamal and

Saeeda Kamal, and their families. As always, their support and partnership were essential elements in my efforts to bring this project to its fruition. My parents, Kamaluddin Ahmad Khan and Sufia Kamal, passed away some years ago. But the values they inspired in me through their own examples, and the encouragement they provided, have been sustaining elements in my commitment to renewable energy. The book embodies those values and is nourished by their encouragement. To them I offer my loving thanks.

Preface
Racing for Survival: Transitioning to a Renewable Energy Path

We are like tenant farmers chopping down the fence around our house for fuel when we should be using nature's inexhaustible sources of energy – sun, wind and tide. I'd put my money on the sun and solar energy. What a source of power! I hope we don't have to wait until oil and coal run out before we tackle that.
    Thomas Alva Edison (1847–1931), 'The father of the electrical age'[1]

We will harness the sun and the wind and the soil to fuel our cars and run our factories ... All this we can do. All this we will do.
    President Barack Obama, Inaugural Address, 20 January 2009[2]

Energy is our basic means of survival. As people everywhere aspire for a better lifestyle, nothing is more convincing than the promise of the better lifestyle that energy can bring. A more comfortable home, good health, eating well, nice clothes, travel, more entertainment, more opportunities to learn – the list goes on. And human ingenuity struck more than gold in detecting, extracting, storing, distributing, using at will and profiting from the energy contained in fossil and nuclear sources. As compact sources of energy, fossil and nuclear sources were amazingly attractive and powerful – like the Genie in Aladdin's lamp, ready to be released and serve its master – without ever getting exhausted.

The Industrial Revolution in the late 18th century was fuelled largely by coal.[3] Since then global industrialization has been accelerated by an increasing reliance on coal and other fossil fuel sources as well, such as oil and natural gas, and uranium for nuclear power, as these were detected over the years. Together and in varying percentages these continue to provide the main fuel sources for the global industrial economy.

The benefits of the fossil fuel/nuclear path need not be discounted. But the story of human triumph with energy, and the benefits it has provided, now has other chapters – some that are newly written, some that were not included, some that were lost and some that were even suppressed. There are problems rooted in this path and, with rapid industrialization, some of these problems have reached a critical level – so much so that, ironically, the very lifestyle that this energy path promised to improve is now threatened by the path itself.

Oil, natural gas, coal and uranium are being quickly depleted. Current scientific estimates suggest that the remaining reserves of oil, natural gas and uranium will be exhausted within the next 50 years, and coal within 250 years.[4] These resources took millions of years to accumulate. We are depleting fossil fuels at a rate 100,000 times faster than they are being replenished, thereby making them nonrenewable.[5]

More urgent is the concern that the crisis is hitting gasoline stations and national defence policies long before the world runs out of oil. We can go on arguing over the exact number of years before the world runs out of oil, and it may not run out of oil completely. But for individual consumers or for economies as a whole, the news is just as bad if they cannot afford it any longer. US oil production already peaked in 1970 and has been declining since. Global oil production is estimated to peak by 2010, triggering an energy crisis causing more price hikes and political tension than we have ever experienced.[6] As the supply of a needed product decreases, the demand increases – unless there is a commensurate decrease in demand or an alternative to the needed product. It's that simple. The skyrocketing oil prices and their worldwide effects, from collapsing economies to wars over control of oil, are symptoms of this crisis. The era of 'cheap oil' has ended. Period. A drop in gasoline prices during an economic downturn or recession makes very little difference. When income drops – due to a lack of pay increases, pay cuts, unemployment – on a massive scale, even the lower gasoline price is no longer cheap for most customers. The long-term trends continue unaffected: downward oil production and upward fuel costs. Stop-gap measures such as the discovery of additional reserves, efficiency and innovation in extraction, and substitution of one nonrenewable fuel for another, at best amount to choosing one disaster over another and keeping the wolf away from the door for a bit longer. We can only go so far while running on empty.

War is not an answer. It's naive to explain any war by a single motive or cause. But it's even more naive to avoid the root cause. History abounds with wars over control of natural resources – land, minerals, water, oil. Nothing better explains the Middle East quagmire than our dependency on, and ever-growing hunger for, oil.[7] The region with two of the largest oil reserves in the world – Saudi Arabia and Iraq – has turned into a bloody battlefield that continues to escalate. About 5,000 US soldiers and an estimated 1 million Iraqis – including innocent civilians – who were killed in the Iraq War are still only a partial toll of the long-term death and destruction caused by the war. When it comes to civilian killing – common to most wars and the warring mentality – any sense of equity and moral threshold seems to have been among the war's first casualties. However it may be justified or politicized, that path of occupation, exploitation, retaliation and destruction will lead us to nowhere but a dead end with more human, environmental and economic costs on all sides.

But let us assume there was an unlimited supply of the conventional fuel resources and nobody needed to go to war over oil. There would still be questions to answer. The entire chain of exploration, extraction, production, utilization and waste disposal is infested with critical, costly and out-of-control health and environmental hazards.

These cause the pollution of air, water, rain and soil – in short, of our ecosphere. We get so accustomed to these consequences that they no longer make news, even if the conditions continue to worsen. Air becomes so polluted that we need to wear a mask; water becomes so contaminated that we cannot go for a swim or fish in a pond or a lake; rain turns so acidic that it destroys vegetation; and soil becomes so contaminated that food grown in it threatens health. Occasional reports or a media blitz sounding a sudden alarm tend to quickly succumb to the desensitizing category of 'old news'. In December 2008 the US news media reported that toxic coal sludge – later estimated to be around a billion gallons – had just spilled over hundreds of acres of land and waterways surrounding the Kingston Fossil Plant of the Tennessee Valley Authority. Further investigation revealed that 'millions of tons of toxic coal ash is piling up in power plant ponds in 32 states, a situation the government has long recognized as a risk to humans and the environment but has done nothing about'.[8] In January 2009, the *New York Times* reported that river pollution from contaminated water from the Alberta oil sand projects in Canada – poised to become the 'new Saudi Arabia' – was causing serious environmental hazards. Susan Casey-Lefkowitz, a senior attorney with the Natural Resources Defense Council in Washington, DC, called it 'one of the most destructive projects on earth'.[9] As we keep entrenching ourselves in the nonrenewable path through conventional and novel means, these and other toxic consequences continue to worsen around the world and their occasional exposures are mere symptoms of a dam of worldwide catastrophes waiting to be broken.

The most talked about consequence of relying on nonrenewables is global warming and its effects on climate change, resulting from carbon dioxide (CO_2), nitrogen oxide and other gaseous emissions from the use of fossil fuels. Global warming to a degree is a natural process which, in fact, is necessary to create and maintain a thermal blanket or an atmospheric shield that keeps the planet's temperature suitable for life. The process is also referred to as the 'Greenhouse Effect' for its similarity to the way a greenhouse nurtures plants by trapping heat inside its boundaries. However, incontrovertible scientific evidence has established that accelerated emissions due to human activities since the Industrial Revolution are causing global warming,[10] acid rain[11] and various kinds of pollution to become environmental threats with critical impacts on the planet's climate and its species. The findings by the world's leading scientists, such as the 130 members of the United Nations (UN) Intergovernmental Panel on Climate Change (IPCC), are no longer refutable by debate. Rajendra K. Pachauri, chairman of the IPCC, warned world leaders at the climate summit meeting at the UN on 22 September 2009: 'Science leaves us no space for inaction now.' What is urgently needed is widespread public education of the findings, challenging political manipulation and denial by vested interests in the conventional energy paths. How much more evidence do we need beyond the growing number of intensifying heatwaves, droughts, melting ice caps and glaciers, rising sea levels, downpours, floods, cold snaps and storms? Even before the catastrophic oil spill in the US Gulf Coast started in April 2010 – spilling millions of gallons to date, Hurricanes Katrina, Gustav, Hanna and Ike caused massive human, environmental and economic devastations in the region; there are more frequent and intense cyclones, typhoons and tornadoes around the world; and drought conditions from California to Australia are destroying food production and fuelling uncontrollable fires. All these things are telling us something about what we have got into by following the nonrenewable energy path. Even worse, they tell us the direction in which we are headed. The catastrophic scenario depicted by the 2004 blockbuster movie *The Day After Tomorrow,* however dramatized and fast-forwarded, could be more than mere fiction. Al Gore's *An Inconvenient Truth* speaks loudly while a California-sized area of ice melts in Antarctica and in the Himalayas the glaciers melt at an accelerated rate. A truth can be inconvenient, but courageously confronting it could be the only turning point away from its consequences.

Nuclear power is often promoted as a solution to climate change. But the promoters carefully avoid mentioning the devastating economic, environmental and political consequences associated with it. To name a few: exclusive and highly secure land requirement, the intensive use of fossil fuels for nuclear power generation (contributing to climate change), radioactive contamination during uranium mining, tilling, processing and power production, massive water usage and contamination, risk of accidents, nuclear waste storage, transportation risks, underground disposal, leakage, potentially devastating impacts of earthquakes and other seismic movements on underground repositories, decommissioning, terrorist attacks on nuclear power plants, nuclear weapons proliferation and out-of-control costs.[12]

The catastrophic consequences of climate change are telling us what we have gotten ourselves into by abusing the atmosphere above ground. With deep geological repositories for nuclear waste we are driving ourselves into an unpredictable and dangerous future by abusing what lies underground.

Since the construction of the first civilian nuclear power plant (Obninsk, with an electrical capacity of 5 megawatts [MW]) near Moscow, Russia, in 1954, there are now 400 nuclear power plants around the world, generating approximately 17 per cent of the world's electricity. While the number of nuclear power plants has increased by 400 per cent, the solution to permanent and safe storage of nuclear waste – which has grown exponentially – has progressed 0 per cent, compounded by all the other associated problems. The waste remains radioactive for millions of years. However it is packaged or bottled, temporary storage of nuclear waste for a few decades or even for a hundred years – in pools, by dumping it in the ocean or by sticking it underground – or passing it on to economically and politically vulnerable communities or nations, are not solutions. When questioned, the nuclear proponents are quick to reply – as they have been doing for years – that scientists are working on a solution. It is like making a toast with a poisoned drink and trying to reassure everyone that chemists are in the lab trying to invent an antidote, although there is no surety that they will ever find one.

Digging for more oil or gas, or excavating another coal or uranium or thorium mine, or building a nuclear 'breeder' reactor, all at an ever greater cost and ever greater risk, are not solutions. Even after the short-term life of a nuclear power plant is over, its radioactive legacy lasts for millions of years and costs billions of dollars. Even if the Gulf Coast oil spill can be stopped, its destructive consequences will continue indefinitely, costing billions of dollars, never truly recovering what has been lost.

The solution requires us to be awakened to the simple truth – and the holistic, perennial wisdom – that the essential condition of sustainability lies in our ability to live harmoniously within the limits and renewability of our natural resources. The age of unlimited, exploitative and imbalanced industrial growth at any cost is over. And the truth should invoke within us an urgent need for a transition – from an obsolete, destructive and unsustainable nonrenewable energy path to a sustainable path of innovation, renewable energy and peace.

Is such a transition possible? The answer is yes, but only under certain conditions. Such a transition is possible only through a worldwide moratorium on further entrenchment into the fossil-nuclear path while, through conservation and efficiency, utilizing these resources only as transitional fuels toward a sustainable renewable energy path. Undoubtedly, the transition will face short-term challenges and risks, but these pale in comparison to a suicidal entrenchment into the dead-end nonrenewable energy path – and that path is inevitable if the risk is not taken. We must set a clear goal and an uncompromising and sustained action-oriented policy for transition. We cannot go on fuelling the nonrenewable path – especially funding it at an astronomically higher rate than funding for the renewable path – and hope that we will somehow be able to reverse the trend. Einstein's saying that we cannot prepare for war and peace at the same time has never been more applicable.

We need practical solutions. However, substituting renewable energy sources for our current nonrenewable fossil fuel and nuclear sources is not as simple as unplugging a toaster from a socket and plugging it into a sunbeam. The transition will not take place without a profound reorganization of social and economic values: it requires transforming our current simplistic conceptions of energy – 'endless', 'cheap', 'bountiful', 'harmless' – into a more sophisticated understanding of the costs, resources, consequences and legacies of our energy choices.

We need a transforming vision of energy in our world and tools to turn that vision into action. We need widespread dissemination of information to understand the worldwide energy dilemma, and to empower people with the tools to tap alternative sources for their energy needs. We need to inspire and motivate people to advocate judicious structuring of energy management in their communities. We need the best expertise in both the nonrenewable and renewable energy paths to unite in a shared vision of sustainability and to lead the transition. What is at stake is our common future – and the common legacy we leave behind for our future generations. We need dynamic plans and programmes of transition to a renewable energy path which are simultaneously tested, strengthened and broadened in their scope through practice.

We need a global perspective. In our increasingly interdependent world, a global perspective is necessary both to comprehend the crisis and find a sustainable solution. Solutions must have global implications so that we do not devise a solution for one place which causes a problem for another. The potential of biofuels does not have to end-up in a 'fuel vs. food' dilemma. The need for transportation or industrial fuel for a few people in one part of the world does not have to be met through massive destruction of the rainforest elsewhere – destroying the biodiversity, ecological balance and the livelihood of those who depend on it. The voracious appetite and unbridled energy consumption by the industrially developed countries – the main cause of climate change – have pushed developing countries and islands around the world to the verge of obliteration under rising sea levels. Sustainability, progress and advancement must be redefined by equity and justice, from a global perspective.

The nature and scope of renewable energy technologies allow these concerns to be addressed in unprecedented ways which are advantageous, both locally and globally.

We need to think holistically. It is futile to consider solutions with a compartmentalized consciousness, and to think without actions. For our actions to be truly effective, both in the present and over time, we will have to base them on holistic thinking that sees the essential connection between fuel depletion, economic recession, environmental destruction and war; and we will have to devise solutions by considering their interconnectedness. Just as these problems are interconnected, so are their solutions. Economic recession and growing unemployment may tend to deter us from thinking afresh and investing in environmental solutions and renewable energy technologies. But that is conventional thinking pursuing economic dead ends. We have to realize that it is only through prioritizing such investments that we will be able to create secure jobs and revitalize the economy. One of the promises of a renewable energy economy is the mutual enhancement of socially and environmentally responsible businesses and economic prosperity.

We will need to think holistically to bridge the conventional gap between philosophy and technology, academics and practicality, development and conservation, economic growth and environment, theory and practice, short-term benefits and long-term solutions, and policy and implementation. Educators, economists, environmentalists, scientists, technologists, business people, media, policymakers and activists – each with much to contribute – will have to engage in a dialogue to devise integrated solutions and actions.

In such actions lies hope. The revolutionary potential of renewable energy technologies offers us myriad opportunities to act on.

The Sun sends an immense amount of energy to the Earth – freely – and it will continue to do so for as long as the life of the Sun, estimated to be between 5 and 10 billion years. Only one hour of sunlight falling on the Earth's surface contains energy equivalent to what we use globally for an entire year.[13] The energy from the Sun, or solar energy, is received through the renewable subsystems of light, heat, wind, water movement and photosynthesis. In addition to direct or natural uses of this energy, we need to technologically convert only a fraction of the solar energy to meet the global energy need. The good news is that such technologies which directly and indirectly capture, convert, store and distribute energy in a wide range of usable forms and scales already exist. The extraordinary variety of renewable energy technologies such as photovoltaics, wind turbines, hydroelectric generators, solar water heaters, solar greenhouses, biogas plants and solar cookers are being applied to a wide range of domestic, industrial and consumer products and purposes. They demonstrate that these practical, ingenious, and economically, environmentally and politically advantageous options are within our reach right now. Such demonstrations can breed success, but only if we promote them. And the best way to promote them is to use them. After all, nothing educates and inspires us more than something that actually works. A range of other options and innovations is emerging on the horizon. Hydrogen fuel cells, biofuels and geothermal options are gaining momentum and maturing through necessary technical, economic, social and environmental scrutinies and criteria for widespread applications. With innovations and breakthroughs, there will be more. But there is a greater possibility for that to happen if we utilize to the fullest what we have at hand to expedite the transition and build a balanced infrastructure that will facilitate future possibilities. We need to lay the transitional path by utilizing the bricks we have at hand.

There is much to be said for the following hopeful developments. Since the shocks of the energy crisis and the Three Mile Island nuclear power plant accident in the 1970s, words such as 'alternative energy', 'solar', 'renewable energy', 'sustainability', 'environment', 'conservation', 'efficiency', 'carbon footprint' and 'green' have found their way into the public vocabulary and consciousness. Books, articles and reports have been published. Conferences, forums and meetings are taking place more frequently than before. Some exceptional legislative and educational developments have taken place. There have been some breakthrough developments in renewable energy technologies, applications and cost reduction. A renewable energy industry was born and it is flourishing. Individuals, international organizations, schools, colleges and universities, sports and recreational facilities, cultural institutions, private industries, governments, non-governmental organizations (NGOs), the United Nations, the World Bank – all of these have initiated a wide range of renewable energy programmes and activities.

But that is not enough. Much more needs to be done, urgently, in every way and everywhere possible to reverse the trend from a collapsing path to a sustainable one. We still fall far short of that. In the face of a growing energy shortage from nonrenewables and escalating energy need – estimated to nearly double by 2030 – barely 7 per cent of the world's energy is generated from renewable energy technologies. Even with all the signs and excitement about going 'green' – encouraging first steps – projections and policies for such generation by 2030 do not exceed 20 per cent. The transition faces other barriers, as well: there are economically and politically powerful vested interests that guard and reinforce our deep entrenchment in the nonrenewable path, often behind a deceptive 'greenwash' of superficial corporate social and environmental responsibility. The investments by these corporations in renewables are a pittance compared to the amount they invest in nonrenewables, while the propaganda for 'clean coal' and 'safe nuclear energy' – each an oxymoron – continues. While people everywhere worry about how to pay for their electricity, heat their homes and water, and drive to work in the face of skyrocketing fuel prices, Big Oil and other energy corporations reap record profits. And they run amok – armed by taxpayers' money and lives – across the globe, under the ocean, and to Antarctica which has been made accessible due to melting ice, to squeeze the Earth of its last drop of oil, the last cubic foot of natural gas, the last pound of coal and the last microgram of uranium. Imagine the risk of the human, economic, environmental and political consequences of this scenario.

On the hopeful side, people around the world are becoming increasingly conscious of the problems – environmental, economic, health, social, philosophical, political, legal, ethical, moral and even spiritual – with dependence on nonrenewables. Al Gore and the IPCC being awarded the 2007 Nobel Prize for Peace is a recognition of the consensus of the global scientific community and the global public concern over the climate-change crisis. It is an urgent call for action, as well. At the same time, the costs of an increasing variety of renewable energy technologies are decreasing, which motivates action. A growing number of incentives such as subsidies, tax benefits and financing options are being offered. Individuals and communities around the world are tapping into these opportunities and acting on solutions. Even countries as a whole, such as Germany, Denmark, Iceland and Cuba, by combining national policies and practical actions are setting into motion an up to 100 per cent transition to the renewable energy path within the foreseeable future. Japan, the largest producer of photovoltaic electricity in the world, has launched 'Solar City' programmes, fully subsidized by the government, to expedite the transition. The Vatican has installed a 2400-module system covering the roof of its main auditorium and is planning a larger solar plant which will feed the surplus energy it generates into the Italian national grid. Abu Dhabi, capital of the United Arab Emirates, built on oil riches, is taking major steps to harness the even more abundant and free renewable resources, sun and wind, to assure a sustainable and prosperous future. It has begun constructing Masdar, a carbon-neutral city mainly relying on renewable energy. The Maldives, an island nation in the Indian Ocean, conscious of its future at risk due to global warming, has announced a plan to become the world's first carbon-neutral country by generating all its electricity with photovoltaics and wind turbines. These are inspiring examples for the world to follow. In the US, Al Gore has announced bold proposals to combat climate change and free the US from its dependency on foreign oil by making a major transition to renewable energy sources within ten years. President Barack Obama, in his inaugural address, promised to 'harness the sun and the winds and the soil', and announced in his later speeches, 'To spark the creation of a clean energy economy, we will double the production of alternative energy in the next three years.'[14] To succeed, these proposals and promises will require fundamental policy changes geared for action and massive public education and participation.

Therefore, there is reason to be hopeful from educating ourselves about the crisis which affects us all and the extraordinary variety of choices which would enable us to act on solutions. Just as our role as energy consumers is universal – however varied that role is – so are our capacities and powers to be more fully responsible and effective choosers and advocates along sustainable paths. We need to nurture such capacities and maximize the variety of choices for translating them into action.

The Renewable Revolution is an invitation to action. It charts a transition from an obsolete, destructive and unsustainable nonrenewable energy path to a renewable energy path of innovation, revitalization, abundance and peace. From a holistic perspective – interweaving technology, economics, science, environment, philosophy, history, spirituality and politics – this book illustrates how we got into an energy crisis and how we can get out of it. This book is educational and practical, brief and comprehensive. It is to be both read and used. Written for a large and crossover readership, this solutions-oriented book should appeal to students, teachers, activists, academics, experts, general readers, socially conscious businesspeople and policymakers – both nationally and internationally. It is a book to unite the crossover readership in a shared vision and diverse actions.

Threaded by the continuum of education to action interwoven in subsequent chapters, in a readable and demystifying way, the book explains the immense potential of the Sun's energy to meet the global energy need; describes renewable energy technologies, such as photovoltaics, wind turbines, hydroelectric generators, solar thermal systems, solar greenhouses, biogas plants and solar cookers, which are available right now and which have the potential to revolutionize the global economy; reports on the current state of development of emerging technologies including fuel cells, biofuels and geothermal systems; show, in words and pictures, exciting examples of renewable energy technologies from a village home in Bangladesh to a skyscraper in New York City; through incisive analysis, debunks the myth and propaganda that renewable energy technologies are too costly and demonstrates instead that one of the real advantages of renewable energy technologies lies with their costs; depicts how, in a variety of ways, individuals, communities, utilities, businesses, educational and cultural institutions, and governments are overcoming barriers and turning the vision of sustainability and peace into action, inspiring more solutions; and finally, puts forth an unprecedented and hopeful Policy-Programme-Practice (PPP) continuum, with both global and local implications.

H. G. Wells put it well: 'Human history becomes more and more a race between education and catastrophe.' Time is of the essence and we all are in the race – and every runner, every step, every contribution counts. The amazingly diverse nature and scope of renewable energy technologies offer us unprecedented opportunities to enter and win the race. It is in our willingness to become more conscious of that race and act upon that consciousness that lie the possibility and hope of turning the direction of the race from catastrophe to sustainability and peace.

1

The Sun: The Enduring Light

The sun, it shines everywhere.   William Shakespeare[1]

Light, as the radiant energy of creation, started the ring-dance of atoms in a diminutive sky, and also the dance of the stars in the vast, lonely theatre of time and space. The planets came out of their bath of fire and basked in the sun for ages.   Rabindranath Tagore[2]

The Sun is our most ancient and sustainable source of energy. A star measuring 1,390,000km across, it is more than a hundred times wider than the Earth. Essentially a gigantic nuclear reactor, the Sun generates tremendous amounts of radiant energy through fusion, which changes hydrogen to helium. From its safe location of 150,000,000km from the Earth, the Sun has been sending energy freely to the Earth day after day, season after season, year after year, for billions of years, sustaining life on this planet. The Sun is the energy source that connects all life. It is the elemental source of our unity, diversity and sustainability.[3] The Sun is the source of the solar system of which we all are a part.

About two-billionths of the Sun's total energy reaches the Earth. That is a tiny percentage of the Sun's total energy, but for the Earth it is an immense amount. It has been estimated that the solar energy that falls annually on the Earth's land surface alone is more than a thousand times the amount of energy now being consumed annually by the entire global economy.[4] According to another estimate: 'In around one hour the amount of sun that hits the face of the Earth is what we use in an entire year, globally, for our energy.'[5]

The good news about the awesome power of the Sun is getting around, beyond the academic research labs and into the industrial sectors engaged in innovating and marketing technologies fuelled by this abundant energy source. Sharp, one of the world's leading manufacturers of photovoltaic modules, took a full-page advertisement in *Time* magazine (15 December 2008) showing a glowing picture of the Sun and a field of photovoltaic arrays, along with the words, 'With the sunlight that hits the Earth in one hour, we could power the world for a year.'

Solon AG, the German solar module manufacturer, contracted the ad agency Jung von Matt to come up with a way to explain this fact so people can actually see it! The result, 'Hail: The Return of the Sun,' is a two-minute film that is unlike anything that one would see in a conventional presentation of scientific data. It begins in a city with a casually dressed man in his thirties sitting down at a typical European open-air cafe table. Just after his drink is served, a battery, about an AA size, plops into the glass from above. Puzzled, he looks up and sees it is just the start of batteries – and more batteries – coming down like hail over the entire landscape. The camera travels from landscapes of North America to Africa to Europe to the Arctic while batteries are showering like hailstorms all over the world. Small batteries and large car batteries keep thundering down to Earth. Windows shatter, cars collide, trucks explode, an igloo collapses, a double-decker bus like the ones in London gets turned over, and people in a frenzy flee for shelter. The sounds are as realistic as they can get. Then the havoc subsides and on a flat dark screen the information appears: '970 trillion kW hours of solar energy hit the earth every day.' Pause. 'Good we can't see it.' Pause. 'Bad we don't use it.' The Sun returns, the landscape brightens, people come out on to the streets looking utterly perplexed, and the camera turns towards a huge field covered with rows and rows of solar photovoltaic panels.[6]

Not all of the energy falling on the Earth can be converted to usable energy, but what makes the scope so promising is that only a fraction of the total energy received needs to be converted to usable energy to satisfy global energy needs. Conversion can be achieved by tapping into the sources through which the energy is channelled, such as light, heat, wind, water movement and photosynthesis. These renewable energy sources supply energy daily, seasonally and annually, replenishing themselves over and over. And the supply of energy is guaranteed for a period as long as the life of the Sun!

However, there is one important factor – even a caution – which requires our attention in order to be able to count on the guaranteed supply of sunshine. It is called 'global dimming', meaning the solar radiation, or the sun's energy, reaching the Earth's surface has been diminishing gradually over time. The sunlight is losing its brightness. This phenomenon was first coined as 'global dimming' by Dr Gerry Stanhill, an English scientist working in Israel. In a paper published in 2001, comparing the record of sunlight during the 1950s through the 1980s, Stanhill found that, while the rate varied by location, worldwide the sunlight was diminishing at the rate of 4 per cent by the time it reached the Earth's surface.[7] His findings confirmed what other scientists have found both before and since.

The potential effect of global dimming would be devastating – for land, water, rainfall and air, including all the life forms who depend on them. Fortunately, scientists have also been able to identify its main cause: pollution due to the burning of coal, oil and wood. Hopeful news is that the attempts to cut down on pollution have reversed the dimming trend and brightened the Earth's surface by about 4 per cent worldwide during the 1990s. Cities across the world, from Los Angeles in the US to Dhaka in Bangladesh, have shown a significant difference in the way their air looks and smells after a period of stricter enforcement of emission standards for automobiles, even though the number of automobiles increased during the same period. Imagine how much more could be achieved simply by reducing the number of automobiles, as well.

That is just what was done to reduce pollution in Beijing during the 2008 Summer Olympics. The Olympics Committee voted to hold the Olympics in Beijing, but raised some concerns over the city's pollution level, especially how it could affect the athletes. A plan of action followed. The Chinese State Environmental Protection Agency restricted the use of private vehicles on the road by banning cars with registrations ending in odd and even numbers on alternate days. The policy resulted in the reduction of the number of cars by half, from the total of 3.2 million per day. The anti-pollution plan also included other measures like reducing or shutting down the operation of some highly polluting factories, promoting public transportation by adding 2,000 buses and 3 new metro lines (with extended operating hours), planting trees, and other 'green' measures including the use of renewable energy technologies. If necessary, even more cars – up to 90 per cent – would be taken off the road during the Olympic days. The dramatically increased visibility due to reduced pollution was highly publicized to convince the world that the city's air quality was suitable for the athletes and the 91,000 visitors. The Olympic Committee was satisfied and the parade began. Now, all these indeed add up to an Olympian feat, deserving our applause!

The message, and warning, of this feat to the world is as significant as any other messages that the Olympics convey. Beijing, a city that only a few years ago invoked the image of clearly visible pedestrians, bikers and sporadic vehicles, had vapourized behind a polluted screen of a car-infested landscape, with all its social, health and environmental consequences. The warning: weigh all sides before rushing into 'Development' and how to fuel it. The warning very much coincides with the Olympics' message of 'Harmony'.

Fortunately, the anti-pollution sideshow seems not to have ended completely with the Olympics. Impressed by the results, China has retained many of the anti-pollution measures in their original or somewhat modified forms. It has passed a rule to ban one in five cars permanently. In July 2009, China even instituted its own version of the Obama administration's 'Cash for Clunkers' programme, offering rebates of $440 to $880 for trading in old high-pollution cars and trucks for new more fuel-efficient vehicles. The policies are continuing to pay off. On 17 October 2009, the *New York Times* reported: 'Beijing's air is actually getting cleaner.'[8] Looking ahead, China has also adopted a plan of becoming a leading producer of hybrid and all-electric vehicles within three years and the world leader in electric cars and buses after that.[9] If China can also make sure that the electricity for these purposes is generated from renewable sources, and not from coal and other nonrenewables, it will be another Olympian feat for the world to witness. China has the renewable energy potential, a rapidly growing renewable energy programme, the technological capability, the political infrastructure and capital to transition to a more sustainable path. The greatest test will be its political will to execute its plan.

What all this adds up to is quite simple: if we want the sunlight to reach the Earth and do us good, we have to stop blocking its path. If we treat the environment responsibly, sunlight can continue to shine brightly and the guarantee of an abundant supply of solar energy will remain in effect. And if we value sustainability, instead of relying on the fossil fuel and nuclear sources which are estimated to be depleted during our children's lifetime or soon thereafter – compounded by the critical and worsening consequences of using them – we can choose to rely on the renewable energy sources which will last at least for the duration of the life of the Sun. The Earth has relied on the Sun for billions of years to sustain itself and its inhabitants, and it can do so for billions of more years. Can we humans – one of the Earth's inhabitants with an awesome mastery of power over its destiny – regain that intrinsic earthly wisdom?

The abundance of energy from renewable sources is a hopeful reminder that the rapidly worsening global 'energy crisis' is not due to an energy shortage, per se. Rather, the crisis is due to the nonrenewable energy path we have chosen. It is a crisis of education, science, economics and politics. It is a moral and philosophical crisis, as well – in the way we look at the Earth and its resources, and our ways of exploiting them. To find a way out of the crisis, we will need to question these premises from a holistic perspective.

Every place on Earth has an immense potential for meeting its energy needs by relying on an appropriate combination of renewable energy sources. If a particular place does not have enough sunlight, it probably has enough wind or hydro power sources; if it does not have enough wind, it probably has enough sunlight, heat or other sources, and so on. Every place has some combination of most, if not all, of the renewable energy sources. And the essential condition of sustainability lies in our ability to live harmoniously within both the limits and renewability of these energy sources – for a nation or for the world community. The promise of renewable energy lies in its diversity, in the amazing variety of natural and technological means through which these energy sources can be utilized. The transition from the energy crisis to a sustainable energy path requires us to expedite that utilization with the utmost urgency. If we can muster the wisdom and the will to do that and act on it, at the end of the tunnel the enduring light from the Sun awaits us – its revolutionary potential practically untapped!

2

Power to the People: Renewable Energy Technologies – Now!

The means may be likened to a seed, the end to a tree; and there is just the same inviolable connection between the means and the end, as there is between the seed and the tree. Mahatma Gandhi[1]

The soft and the hard energy paths, and a myriad variations on their themes, appear to be the only choices there are, and we must decide which we prefer. Amory B. Lovins[2]

Imagine a complete power plant that's small enough to electrify a rural home or large enough to electrify an entire community. Imagine a power plant that not only generates electricity for your home but, if connected to the power line, actually 'turns the meter backward', making you not only a producer of electricity from your own power plant, but also a seller of electricity to your utility. Imagine that the technologies that operate these power plants will never run out of fuel as long as the Sun is shining, will not pollute or cause climate change, and will not require you to look beyond your rooftop or backyard to set them up and plug them into their power source – the Sun!

To most people that kind of imagining may sound like wishful thinking, but it is, in fact, the welcome description of the revolutionary, hopeful and ingenious world of renewable energy technologies – not in the future – but now!

Taken against the backdrop of today's energy crisis – the deeply troubling combination of growing fuel scarcity, skyrocketing prices, economic recession, global warming, environmental destruction, collapsing centralized power plants and energy wars – the renewable energy field is teeming with innovative possibilities. And the extraordinary variety of technologies which are already available dispels the myth that renewable energy technology is purely an exotic fantasy for the future. Many of these economically, environmentally and politically advantageous options are within our reach right now.

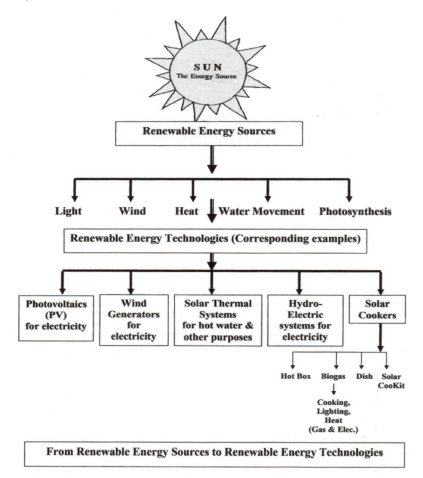

Figure 2.1 *From renewable energy sources to renewable energy technologies*

A global survey of renewable energy projects and products provides a glimpse into the widespread applicability of these technologies. Ranging from microwatt to megawatt scales, applications already range from space satellites to terrestrial applications in lighting, heating, cooling, water pumping, food storage, medical refrigeration, televisions, computers, calculators, toys, microwave transmitters, cooking, environmental monitoring, aviation aids, traffic control systems, telecommunication, remote sensing, marine navigation, printing, desalination, cathodic protection, dairy machineries, boats, cars and other means of transportation, and a wide range of other domestic, industrial and consumer products and purposes. Practically every country in the world is implementing one or more types of renewable energy technologies.

What are these technologies? Here are some basic explanations. Beyond the science and technology part, there are some interesting stories behind them, too. If you are new to this field, that's fine. These explanations are not expected to turn you into a renewable energy expert – but, then again, you may follow up and actually become one! The purpose here, however, is more immediate and practical. It is to generate enough understanding about these technologies and demystify them so you can see that much of the revolutionary potential of renewable energy technologies for a transition to the renewable energy path truly lies with people – with individuals, groups, communities and so on, whether by installing their own renewable energy systems or by putting informed pressure on the policymakers to produce and use energy from renewable sources. It is a transition, in a very practical, participatory and political sense, to the power of the people. With that understanding – and self-empowerment – hopefully you will then be able to realistically assess how you can contribute to the transition, and feel equipped and inspired to act!

Photovoltaics

Photovoltaic cells (PV) convert light directly into electricity. The cells are manu-factured from semiconductor materials such as silicon, one of the most abundant elements on Earth. Most commonly silicon is found as specks of sparkling materials in sand. When sunlight falls on silicon, it generates electricity – naturally. Under laboratory conditions, silicon is grown as cylindrical crystals, which are then thinly sliced and finished to create solar cells. A PV panel is a sealed unit that contains an interconnected series of solar cells and produces a rated voltage and current. Typically, the panels produce between 12 and 18 of volts direct current (DC). If necessary, the electricity can be inverted to a different voltage and alternating current (AC). In addition to the single-crystal silicon type panels, there are also other types, such as polycrystalline, thin film and amorphous silicon panels.

The electricity-generating capacity of a PV panel depends on six basic factors: the intensity of sunlight, the duration of sunlight, the size of the panel, the efficiency of the cells to convert light into electricity, the direction of the panel and the tilt angle at which the panel is placed to face sunlight. A panel, or a set of panels, can be placed on a fixed mount at an angle where it gets the maximum sunlight throughout the day, or on a flexible mount adjusted seasonally to maximize the annual reception of sunlight. It can be placed vertically on a wall, flatly on a roof, fixed on a pole, or on a pole-mounted tracker that follows the path of the Sun throughout the day – like a sunflower – to maximize the energy production.

It has been estimated that the sunlight at noon striking an area 112.5km long by 112.5km wide, if converted to electricity through photovoltaics, would equal the peak capacity of all the Earth's power plants. A photovoltaic spread equivalent to only 1 per cent of the Sahara Desert would produce all the electricity consumed on the planet.[3] But we do not need to build one centralized power plant of that size to meet the global energy need. Rather, the revolutionary potential of photovoltaics can be realized through a multitude of scales and designs. PV systems can be large, multi-megawatt, centralized power plants in different parts of the world – appropriately located in fields and deserts.[4] And they can be small systems, installed on the roof or the side wall or the yard, enough to meet just the energy needs of one single-family home or even a portion of the home's needs.

Scientists are also exploring the feasibility of building 'Solar Power Satellites' or, as Japan calls it, 'Space Solar Power System' (SSPS), each consisting of a large photovoltaic power generator and transmission panel, which will collect energy in space and zap it down to Earth, using laser beams or microwaves, to be collected by gigantic parabolic antennae. There are scientific, technical, economic, environmental and political issues still to be resolved.[5] In the meantime, take this as another glimpse of the enormous potential and the multitude of ways that PV can be utilized. We just have to tap into it sensibly and responsibly. Fortunately, much of the potential is already realizable – in 'down-to-Earth' ways!

In varying amounts, all regions around the world receive sunlight. And panels can be installed on roofs and walls and yards that already exist, on rural homes or urban skyscrapers, covering them fully or partially, as long as sunlight reaches them. This makes photovoltaics a universally appealing and appropriate renewable energy technology.

Modularity

A PV panel is also called a 'module' because of the flexibility it allows in sizing a system. A complete PV system can be designed with just one module that is connected to a load and an electricity storage unit, or with many modules. Modules range in size and their rated capacities of power production. They can be under 10 watts for charging small items like solar lamps and battery chargers, going up to 200 watts or more for larger systems. They can be custom produced in size and shape according to particular design criteria. The system can also be designed on a larger – even megawatts – scale by adding more modules and other system components. A unit of more than one interconnected modules is called a 'solar array'.

Modularity allows a PV system to be designed to fit individual energy needs and budgets. If energy needs increase, the system can be enlarged. Modularity is at the heart of the basic simplicity and versatility of PV technology, and it has tremendous implications for the cost-effective transfer, application and popularization of the technology around the world.

Versatility

Photovoltaic systems can be designed as four primary types: stand-alone, hybrid, grid-connected and utility-scale.

Stand-alone

A stand-alone PV system is a self-contained system with one or more modules, light fixtures, a charge controller and a battery (or batteries). It can be stationary or portable, and it serves as an instant power plant where it is installed. Exact prices vary according to where a system is bought, the choice of system components and the volume of purchase. On average, a 50-watt system currently costs around $450; the module alone costs around $300. One 50-watt module placed in the sun for five hours will generate enough electricity to power a 250-watt incandescent light bulb for five hours. Replacing the incandescent bulb with a fluorescent bulb or tube, which are more efficient, could nearly double the energy use. Light-emitting diode (LED) lights, which are gaining popularity with solar system manufacturers and users as their costs are coming down, can add even more hours to a system. If energy needs and affordability increase, the system can be enlarged with minimal lead time. Hundreds of thousands of 40-watt one-module systems – each of which can power three 8-watt fluorescent tubes for about eight hours every night – have been installed in rural homes, schools, clinics, shops and community centres around the world. Where upfront payment of the full cost of the system is an obstacle, microcredit and other financing options have facilitated the rapidly growing popularity of PV usage.

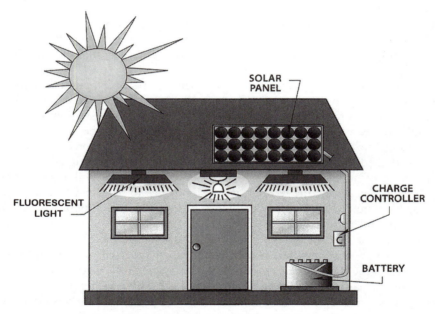

Stand-Alone Photovoltaic (PV) System

Figure 2.2 *Stand-alone photovoltaic (PV) system*

The photovoltaic module converts sunlight directly into electricity. Electricity flows through a charge controller (C) into a battery (12V) to be stored and then distributed for lighting and other purposes. One 50-watt module will produce about 250 watt-hrs/day of electricity with the daily availability of 5 hour-peak sun/yearly average. Larger systems can be designed with additional modules and other components. A PV system can be stand-alone (as above), hybrid (combining PV and non-PV, such as a wind turbine), grid-connected (connected to the utility grid which provides both the storage and back-up power), and utility-scale (large central power plants connected to the grid).

Hybrid

A hybrid PV system combines photovoltaic technology with a non-PV backup or supplemental power source to complete the system. Together, the two types of technology share the load of supplying the required energy. The backup could be another renewable energy technology, such as a wind turbine or a hydroelectric generator, or it could be a conventional generator fuelled by gasoline, diesel or propane technology. Systems that combine PV technology with conventional generators are called 'PV/Genset hybrid systems'.

Grid-connected

A grid-connected PV system feeds the power generated by the PV system into the utility grid, which provides both the storage and backup power for the system. The system requires an inverter to match the PV-generated DC power and frequency to the AC power and frequency of the utility grid. The system can have its own emergency backup power stored in a battery bank in case of a power outage in the utility grid. The electric meter tracks the amount of power the PV system feeds into the utility grid. This type of system is also called a 'grid-tied' or 'utility-interactive hybrid' system.

The modules can be installed on the roof, wall or a space adjacent to the building for which it will generate electricity. They can also be installed as part of the building materials, such as roof tiles, slates, shingles or a 'PV skin' on the wall, in which the modules replace the usual building materials. In these cases, the system is called a 'building-integrated system'. An advanced generation of PV comes in the form of durable and transparent glazed glass, allowing light to pass through, with the appearance of a skylight, glass wall or window glass. These building-integrated options not only make the technology least obtrusive to the building design, but also add a combined aesthetic-utility dimension to it. PV is revolutionizing building technology and architecture.

The current cost for grid-connected systems ranges from $7 to $9 per watt installed, before any subsidies, rebates, tax breaks or other financial incentives are taken into account. The price range reflects differences in the place and volume of purchase. The cost has been coming down over the years. With growing demand, mass production and a more efficient manufacturing process, the downward trend is likely to continue. Grid-connected systems with PV roof tiles or shingles cost more, about $12 per watt, but they save money by eliminating the cost of conventional roofing materials. One of the revolutionary advantages of all the grid-connected systems is that the owner of any such system is not only a customer of the utility but also a seller of power to the utility, according to the amount of power produced by the system. Once the system is connected, the payback begins instantly. This is sometimes referred to as 'turning the meter backward'.

An average three-bedroom American home will require a 5-kilowatt system to meet its electricity needs. At around $8 per watt, installed, a system of this size carries a price tag of $40,000. If this figure looks intimidating or uninviting, understanding that this is a one-time payment for electricity for many years might make it more palatable. With free fuel being supplied by the Sun, the cost curve of this system

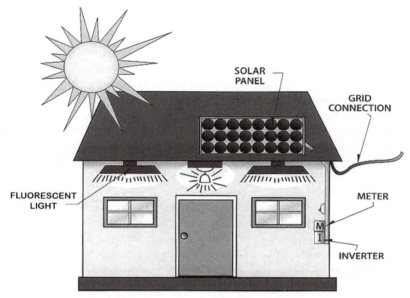

Grid-Connected Photovoltaic (PV) System

Figure 2.3 *Grid-connected photovoltaic system*

The photovoltaic module (PV) converts sunlight directly into electricity. Electricity flows through an inverter (I) that changes DC power and frequency generated by PV to AC power and frequency of the grid electricity. Meter (M) keeps account of the electricity being drawn from the grid or fed into the grid from PV. The utility grid functions both as the storage and back-up power. Larger systems can be designed with additional modules and other components.

decreases over time. Compare that to the cost of electricity from nonrenewable sources in the future, particularly if the currently skyrocketing energy prices and growing fuel scarcity are any indication. Subsidies, rebates, tax breaks and other incentives, and more affordable payment plans offered by the federal and state governments, utility companies, environmental organizations, foundations and banks also serve to offset initial costs and can underwrite up to 50 per cent of the system cost, upfront and over time. Sellers of solar systems usually include this information in their marketing packages and bids. They also handle the paperwork related to obtaining the upfront subsidies, simplifying the transaction for the customer who may otherwise find the process unfamiliar, if not intimidating.

The modular advantage of the technology allows for a system to be sized to the customer's current budget. For example, the installation of a 2-kilowatt system, instead of the full 5-kilowatt system. Systems can be installed to meet partial electricity consumption needs, regardless of whether the demands are from a stand-alone home or a multi-unit building, such as a condominium or a co-op. Systems can also be designed to power the common area of the building.

Utility-scale

A utility-scale PV system is a large central power plant that is connected to the utility grid. In this type of system, a large array, or a set of arrays, feeds power directly into the utility grid, which then distributes power through conventional power lines to its consumers. Gaining popularity, this arrangement has been well-tested over the years. An early example, the utility-scale Carissa Plains plant in California – the largest of this type at the time – was spread over dozens of acres of land. It was built by ARCO Solar Corporation in 1984 to generate 6.5 megawatts of electricity to feed the grid owned and operated by the Pacific Gas & Electric Company.[6] The experimental plant, generating enough electricity for 2300 homes, is now dismantled, but provided much valuable data for system types and material development that are benefiting other utility-scale systems elsewhere.

The new millennium dawned with the surging of construction of large utility-scale PV power plants. Two of these are located in Germany. The Bavaria Solar Park, a 10-megawatt plant in Mühlhausen, Bavaria, with three solar plants adding up to its total capacity and with a PV surface area of 62 acres, generates enough electricity to power 9000 homes. Operating since 2004, it was built by the Berkeley-based PowerLight Corporation (a wholly owned subsidiary of SunPower Corporation) in collaboration with the German government.[7] The second one, the Pocking Solar Park in the Lower Bavarian town of Pocking, with a peak performance of 10 megawatts, became the largest one-piece photovoltaic power plant in the world. Costing 125 million euros, with construction and assembly starting in August 2005 and inaugurated on 27 April 2006, the Pocking Solar Park is a result of a joint effort by Martin Bucher Projektentwicklungen (Stuttgart), Shell Solar GmbH (Munich) and CommerzLeasing und Immobilien AG (Düsseldorf), with the proud support of the people of Pocking. An inspiring transformational feature of this solar park, with modules covering 19 acres, is that it is built – after extensive tests and surveys – on a former 79-acre military training ground which also serves as the rearing ground for 600 ewes and 200 lambs, which graze the green pasture under and around the PV arrays![8] That, I think, expresses well a universal meaning of the biblical saying that 'you shall beat your swords into ploughshares and your spears into pruning hooks'.

An 11-megawatt utility-scale photovoltaic power plant began construction in May 2006 and was completed in a year in Serpa, 124 miles southeast of Lisbon, Portugal. The 60-acre photovoltaic surface area has 52,000 modules mounted on PowerLight's solar tracker system which follows the sun throughout the day, generating more electricity than conventional fixed-mount systems. The principal project partners are Cataveno, a leading Portuguese renewable energy company, PowerLight Corporation and GE Energy Financial Services, which will own and finance the facility in a $75 million transaction. The southern-facing hillside under the plant will remain productive farmland, generate enough electricity to power 8000 homes and farms, and will reduce greenhouse gas emissions by more than 30,000 tons a year.[9]

The period of 2007–9 literally saw an explosion of utility-scale PV power plants coming on line. To date, there are at least 1000 large (peak power > 1MWp) utility-scale PV power plants in the world. The plants are mainly concentrated in Spain, the lead country, and Germany, with others located in South Korea, Thailand, China, Japan, Czech Republic, United Arab Emirates, Portugal, Italy, Switzerland, France, Belgium, Canada and the U.S. The top 50 power plants range between 11.45MW to 60MW.[10] The trend continues. Even larger plants are in the works.

Reliability

Photovoltaic technology is exceptionally well-tested and has been in existence for more than half a century. The technology has its origins in the early research and development of the US space programme. The research came about to solve an acute problem: satellites needed a highly reliable, lightweight, durable energy device that could also be fuelled locally. Photovoltaic technology proved to be an ideal solution. *Vanguard 1*, a satellite powered by a 108-cell PV system, was launched into orbit by the United States on 17 March 1958. The results exceeded expectations. Not only did the system perform reliably, it also outlasted the satellite itself! Since then the technology has continued to undergo breakthrough improvements in cell efficiency, scale of production, manufacturing process and cost-effectiveness, making it increasingly viable for widespread terrestrial applications.

Seasonal variations can make a significant difference in electricity generation from a particular system. The generation during winter can be down by as much as one-third to one-half of the summer production when sunlight hours are greater. That is an important consideration in sizing a stand-alone system, with no grid back-up. So the choice is to design a system large enough to reliably supply enough electricity to meet the demand year-round, or a smaller system that can supply plenty of electricity during the summer but will require the user to be more conservative in electricity usage during the winter. Ultimately, it is up to the user to decide about the size of the system, taking the seasonal variations for generating electricity and its usage into consideration. And if one can and does choose to invest in a system that is large enough to match the winter demands, therefore having an energy surplus during the summer, well, it may just be what the user needed to run some power tools to finish a project that has been lying dormant during the winter! The point is that a stand-alone PV system can be designed in consideration of the effect of seasonal variations on electricity generation. By designing it with an adequate module size and storage capacity, it can deliver a reliable amount of electricity year-round.

Durability

A PV module undergoes rigorous testing for its durability against every conceivable type of weather – sunlight, rain, wind, hailstorms, temperature fluctuations, snow – before it is released to the market. Field tests have proved that modules reliably generate electricity for 40 years or more without a significant loss of efficiency. Current modules are available with a warranty of up to 25 years, up from the average warranty period of 10 years only a decade ago. With field test results and the continually improving manufacturing process, it is likely that future modules will be available with longer life warranties and an even longer actual life.

Photovoltaic modules contain no moving parts, which is one of the best safeguards against malfunctions or breakdowns. And the basic simplicity and user-friendliness inherent in a PV system contribute significantly to its proper maintenance, and thereby also its durability. Battery monitoring and maintenance can be learned easily. The maintenance need for the module itself is minimal: wiping with a wet cloth every few months does it. And, of course, Mother Nature can be credited with doing a lot of the cleaning – for free – every time it rains!

Durability must come from the whole system, not just from some of its components. Fortunately, components other than the PV module, such as the charge controller and battery (for a stand-alone system) and the inverter (for stand-alone and grid-connected systems), have gone through some breakthrough improvements over the years. The charge controller used to be the most trouble-plagued component not too long ago. But this is not so any more. A growing variety of charge controllers available today are more durable, efficient, compact and cheaper than their predecessors.

Batteries have come a long way, too. A growing variety of maintenance free, sealed, deep cycle, less toxic, long-lasting and cost-effective batteries are available. To attain durability, environmental safety and economy, encouraging the use of such batteries must be a priority. Used batteries must be disposed of safely. Some communities conduct 'Household Hazardous Waste Collection Days'. Some battery manufacturers offer environmentally responsible battery disposal programmes. The American Automobile Association (AAA), a 50-million member not-for profit automobile lobby group, service organization and seller of vehicle insurance, offers battery recycling options to its members. Some sellers of PV systems are beginning to take such initiatives. It is highly commendable that in 2004 Rahimafrooz Batteries, the largest seller of batteries for cars and solar systems in Bangladesh, became one of the first such businesses in the world to institute a recycling programme for used batteries. The programme even includes free pick-up and part of the proceeds from the recycling are used for environmental welfare.

As the use of stand-alone PV systems spreads, battery recycling must become an industry-wide regulation, as an integral part of battery sale and disposal. Solar-powered fuel hydrogen cells are on the horizon, which will add a revolutionary dimension to battery technology in terms of its increased capacity for energy storage and environmental advantages.

The combination of its modularity, versatility, reliability and durability – fuelled by free and abundant energy – makes photovoltaics a technology whose future is as bright as sunlight! The news is getting around, from the remotest parts of the world to the urban metropolis. It shows in the dramatically rising demand for PV in recent years. By 2009, the cumulative global PV installations reached over 15,000 megawatts. Doubling every two years since 2002, it is the fastest growing energy technology in the world.

Wind turbines

Wind is the movement of air caused by the Sun's uneven heating of the Earth's land and ocean surfaces. This uneven heating causes the air over the warm surface to expand and rise, and then to move to let the heavier colder air take its place, causing wind. Throughout the centuries, and throughout the world, this natural, renewable power source has been tapped to perform many tasks. Wind has powered sailing ships and boats, and turned windmills to pump water, grind corn and chalk, crush seeds and saw wood. To this day, especially in the Far East, many boats carrying people or cargoes continue to sail down rivers at an unhurried pace, silently and gracefully. In the West, sail boating and sail fishing are enduring popular sports, powered by wind.

Windmills, used as early as 1300 years ago in Persia, can still be seen in many parts of the world today. One of the largest concentrations of windmills today exists in The Netherlands, where about a thousand wooden windmills scattered throughout the countryside stand witness to a tradition of building windmills in Europe that began about 800 years ago. There is also the legacy of Cervantes: Don Quixote charging the windmill – a witness to its time and a timeless treasure in our literary heritage.

Following World War I, some farmers in the U.S. came up with a novel use of army surplus aircraft propellers by connecting them to simple electric generators – and the widespread generation of electricity with windmills, or 'wind turbines', began.[11] Between 1928 and 1960, the American classic 'Jacobs' wind turbine, named after its inventor, Marcellus Jacobs of Minnesota, popularized the technology in the U.S. and in some other countries. Unfortunately, the rapid expansion of the centralized power lines of the 'cheap fuel era' not only brought a virtual end to the further development of wind turbines, it actually displaced them. The Jacobs became history – except for a few mavericks like the Texan farmer Joseph Spinhirne, who purchased his Jacobs for under $100 in 1947 and kept using it past the 'cheap fuel era' with a clear sense of conviction and foresight that its time was to come.[12]

Especially since the oil shock of the 1970s, the generation of electricity by wind turbines has been experiencing a 'second wind'. Even through the 1980s – a decade which saw declining public support for alternative energy issues and sustainable energy policies – wind-turbine technology continued to develop into more efficient, reliable and cost-effective products. The US, especially California, and several European countries, notably Denmark, Germany and Spain, are to be credited for this resurgence. In California, between 1980–90, the number of wind turbines rose from about 150 to over 18,000, generating almost 2 billion kilowatt-hours of electricity, equivalent to 3 million barrels of oil and enough electricity for 300,000 homes.[13] By 1988, the turbines installed in Altamont, Tehachapi and San Gorgonio passes were generating 1.827 billion kilowatt-hours of electricity, equivalent to about 1 per cent of California's demand and as much electricity as San Francisco consumes in a year.[14] In Denmark, more than 6000 wind turbines provide 20 per cent of the country's total electricity consumption.[15]

Wind turbines come in a wide range of sizes and capacities. A small 250-watt model can power a cabin or a small rural home. At the other end of the spectrum, with turbines of many in-between capacities, some manufacturers are working on wind turbines with capacities over 5 megawatts each. Since acquiring Enron's wind power division, GE has already introduced 3.2 and 3.6 megawatt models. One 3.6 megawatt turbine has the capacity to power about 1500 homes.

Wind turbines can be installed as stand-alone, hybrid, grid-connected or utility-scale systems. A series of wind turbines can be installed as a 'wind farm' to create a utility-scale power generating system, such as the one at Altamont Pass in California.

Five main concerns with the first generation of wind turbines have been the frequency of breakdowns due to moving parts; the range of wind speed within which a wind turbine can operate; noise; land requirement; and aesthetics. The current generation of wind turbines demonstrates vast improvements over previous models. They contain fewer moving parts, are made of more durable materials and are installed with an expectation of reliable operation for 30 years or more. The operating range has broadened from around 24–97kmh to 10–193kmh wind speed. Wind does not blow at the optimum speed all the time, but a low cut-in speed helps generating electricity by wind blowing much of the time. A taller pole on the ground, or even a small pole mounted on a tall building, can add significantly to the amount of electricity generation. Design improvements make them more efficient and less noisy. In fact, one line of wind turbines has been named 'Whisper'!

Land requirement can be a concern with wind farms, which should be properly sited. Uninhabited sections of coastal belts, islands, deserts and mountains are ideal sites. Land use for wind farms does not need to be exclusive. Because the turbines rest well above ground level, much of the land in a wind farm is usable for other purposes. A wind farm can be integrated with an agricultural farm. This has been done in the US, Germany, the UK, The Netherlands and India, setting examples of growing food and generating electricity – which also means additional income – simultaneously! The trend is growing, globally, for both large and small farms.

Concerns have also been raised about the effects of wind turbines on birds. Especially if they are installed blocking the birds' flight routes, birds can be injured or even killed. However, smog and oil spills are also lethal to birds, so wind turbines should not be singled out as anti-wildlife. No other organization is in a better position to speak on this than the National Audubon Society, with its mission to 'conserve and restore ecosystems, focusing on birds and other wildlife for the benefit of humanity'. The mission encompasses its advocacy for renewable energy, as well. It therefore supports wind turbine projects through close attention to the concern for their impact on birds. The concern has also contributed to installing wind turbines with ample space between, preserving safe flight routes for birds. Siting and designing a wind farm, therefore, must go hand in hand.

Innovative ideas for siting wind turbines continue to emerge. Wind turbines can be interwoven in cities and semi-urban landscapes with ample wind. Adjacent tall buildings in cities and semi-urban landscapes often create wind tunnels between them. The World Trade Center of Bahrain, a kingdom on the Persian Gulf, sets an example. The 50-floor, 240-metre high, twin-tower complex has installed three wind turbines on each of the bridges connecting the buildings. Inaugurated in April 2008, the three turbines of 225 kilowatt capacity each are expected to provide up to 15 per cent of the towers' electricity consumption. In the US, 'Windy City' Chicago, and other cities with tall buildings and wind, take notice!

In Bangladesh, on the roof of a cyclone shelter in the coastal region of Pekua, a small 300-watt stand-alone wind turbine system was installed in 1998 to generate electricity for some of the rooms of the shelter. The project was part of the 'Solar Energy Program for Sustainable Development' of BRAC, an NGO originally known by the fuller title Bangladesh Rural Advancement Committee, which built and maintains the shelter. The cement-built four-floor structure has very few windows to prevent sea waves from breaking into the building in case of a cyclone. But that also prevents light from entering the building. The wind turbine – installed with a 3m metal pole on the shelter's windy roof – became an ideal way to provide electricity to meet some of the lighting needs of the shelter, not only during cyclone seasons, but also year round in this generally windy region, enhancing the shelter's use for various other purposes.

In Boston, Logan International Airport became the first commercial airport to generate electricity with wind turbines. In July 2008, 20 utility-connected wind turbines were installed onto the parapet of the Logan Office Center. Architecturally designed and installed by California-based AeroVironment, Inc., each with a rated capacity of 1 kilowatt, the turbines are expected to generate about 100,000 kilowatt hours of electricity annually, equal to about 2 to 3 per cent of the building's energy needs. The $140,000 project is expected to save between $12,000 and $15,000 annually on Logan's energy bill. That is an attractive ten-year payback period, especially because the turbines can be expected to keep generating electricity with free fuel for a warranted period of 20 years or more.[16]

Offshore siting of wind turbines is another major option. Thanks especially to the leadership of the United Kingdom, Denmark and Germany, this option is gaining popularity around the world. According to the European Wind Energy Association, in 2009 the total capacity of offshore wind power plants in Europe has reached 2056MW. Eight new power plants came into operation in 2009. An additional 1000MW of offshore wind energy is projected for 2010.[17]

The first offshore wind farm in the US is slated to be built on Horseshoe Shoal in Nantucket Sound, five miles off the Cape Cod shore in Massachusetts. Cape Wind Associates and its managing partner Energy Management Inc. (EMI) plan to put the project into operation. It will consist of 130 wind turbines, 3.6MW each, spaced about one-third of a mile apart and connected by undersea cables, expected to be completed in 2012. The 'wind park', as the project owners call it, will not only be nearly invisible on all but the clearest days and will have little effect on fishermen and recreational users, but it may also be a tourist attraction as well.

The Cape Wind turbines will generate around 450MW of electricity, enough to power more than half a million homes. The project has undergone extensive environmental, geological, meteorological and engineering studies. Its approval rate has gone up significantly in polls conducted nationally and an increasing number of organizations are supporting it, including the Massachusetts Audubon Society. But it is not without opposition. The main objection comes from some Cape residents who object to it on the grounds that it interferes with the ocean view. Politics, too, seems to play a role here.

So, there are also the aesthetic and political concerns to address. How one judges an open landscape or ocean view interrupted by a single wind turbine or an entire wind farm is relative to the eyes and values of the beholder. And the judgement can rest at either extreme end of a scale. It is not the first time that such concerns have been raised, and it will not be the last. On that ground alone, the objections to the Cape Wind project should not come as a surprise. But, guided by a fair and equitable policy, the projected savings of $800 million over 20 years, the creation of jobs, revenue from tourism and the elimination of 4624 tons of sulphur dioxide, 120 tons of carbon monoxide, 1566 tons of nitrous oxides, more than half a million tons of greenhouse gases and 448 tons of particulates from being dumped into the air – topped by energy independence through an unlimited and free renewable energy source – certainly could, and should, be reasons to sway the wind toward a favourable direction.[18]

Innovations in the design of turbine themselves are adding a whole other look to the technology. Vertical axis wind turbines – in contrast to the conventional horizontal turbines with sword-like blades – are gaining popularity. These can have a look of a vertically placed barrel or a cone with openings in the body which catches wind from all directions while rotating on a base or on top of a pole; a drum cut in halves, joined in the middle, vertically placed, facing in opposite directions and rotating; two seashell-like wings placed and operating the same way as the drum; and with bow-shaped blades with one end attached to the top and the other to the bottom of the pole, which rotates with wind from any direction, powering the turbine.

The idea of the vertical axis turbine is not new. Way back in the 1970s, threatened by the Arab oil embargo which prompted the 'alternative energy movement', some of these designs were conceived and highly promising experimental models were installed in the U.S. Unfortunately, while the wind turbine industry in Europe continued to experience a more favourable climate, in the US, as with the movement itself, soon the politics of nonrenewables had the upper hand, support for innovation declined and the potential of vertical axis turbines got shelved. Fortunately, steeped in that history and building on that foundation, vertical axis turbines are making a come back.

One of the most exciting examples is the Greenpower Utility System (GUS) line of vertical axis wind turbines. This is available in a range of sizes. Designed with a twin spiral and vertical blades, it has a sculptural appearance. Its other attractive features are that: it is highly efficient – producing up to 50 per cent more electricity on an annual basis versus conventional turbines with the same area; it has a high operating range with low cut-in speed – generating electricity in winds as low as 6.5kmh (1.5m/s) and continuing to generate power in wind speeds up to 210kmh (60m/s), depending on the model; it is highly durable – withstanding extreme weather such as frost, ice, sand, humidity and wind conditions greater than 210kmh (60m/s) (it has been tested on the snow-covered top of Mt Washington!); it is safe – with a compact body spanning a small diameter area, it is less prone to block birds' or other flying animals' passage; and it is practically soundless. What is also especially attractive about GUS is that, in contrast to the usual steel grey exterior of most wind turbines, the manufacturer will decorate it with custom colours and graphics embedded onto the blades. That, I think is a brilliant way to add a little colour to your wind![19]

In areas with consistently sufficient wind, wind turbines are proving to be the least expensive and most appropriate electricity-generating technology. The technology is growing at a record-setting rate. According to the Global Wind Energy Council, worldwide the installed wind capacity in 2008 had reached 94,123MW, averaging an annual growth rate of almost 30 per cent over the previous ten years.[20] In 1995, it was less than 5000MW. The top five countries leading the trend are Germany, the US, Spain, India and China. Other countries, including the Netherlands, Belgium, Portugal, Finland, Italy, Greece, Brazil, the UK and Ireland, are expanding their wind energy programmes.[21]

The answer – at least a good part of it – 'is blowing in the wind'!

Hydroelectric systems

Falling water turns a turbine to generate electricity. This simple principle gave rise to the massive development and implementation of hydroelectric technology, making it the source of 20 per cent of the world's total electricity today. Still, its full potential remains untapped. However, hydroelectric technology has some cautionary lessons to teach even when it comes to using renewable energy. It is a special reminder why we must assess each technology – renewable or nonrenewable – holistically to decide which ones to utilize, and how and to what extent we do this.

Most hydroelectricity is generated by constructing a dam across rivers or lakes. For years, the major trend has been the bigger the better, the largest one being a 10-gigawatt hydroelectric dam in Venezuela, completed in 1986. An even larger system of 12-gigawatts size is under construction in Brazil.[22] The world's largest hydroelectric dam is under construction in China. Yet, a lesson to be learned here is that large hydroelectric dams cause serious ecological damage. Some of the damages have been summarized by Michael Brower as follows:

> *Depending on the location, reservoirs created by large dams inundate forests, farmland, wildlife habitats, and scenic areas. In addition, dams cause radical changes in river ecosystems. Sediments bearing nutrients accumulate in reservoirs instead of being carried downstream, and changes in river-flow rate, temperature, and oxygen content alter the balance of plant and fish life. Reservoir evaporation increases salt and mineral content, a serious problem ... Dams also block fish migration and destroy spawning grounds, resulting in losses for commercial fisheries and sport fishing.[23]*

Furthermore, John Naar has noted, 'In tropical regions reservoirs act as breeding grounds for carriers of malaria, schistosomiasis, and river blindness.'[24] Hydroelectricity generation is also vulnerable to drought, as dramatized by the crisis faced by Egypt's Aswan High Dam[25] and the dams in Brazil.[26]

Worldwide, large hydroelectric dams have forced massive displacements and irreparable damages to people – often politically marginalized indigenous communities. Such damages have also been caused, and continue to be caused, by other large and centralized nonrenewable power plants fuelled by coal, oil, natural gas and uranium. All too often indigenous people's desperate outcries get washed aside by the lopsided march of 'development', insatiable hunger for energy, and lure of profits for the few and powerful. It is against such destructive consequences that opposition to the Hydro-Quebec in Canada, the Kinzua Dam on the upper Allegheny River in Pennsylvania, the Mahaveli Dam in Sri Lanka, the Narmada Dam project of a series of large hydroelectric dams on the Narmada River in India, the Kaptai Dam project on Karnaphuli River in the Chittagong Hill Tracts area in Bangladesh, the proposed Tipaimukh Dam project in Manipur State of India, and other large-scale hydroelectric dams has surfaced. Victimization – amounting to cultural genocide – still continues, while only a few protesting voices, such as the people-centred 'Narmada Bachao' (Save Narmada) movement, are succeeding in internationalizing this critical issue and bringing it to the world's attention.[27] Through joint opposition from environmentalists, geologists, grassroots organizations and other activists both in India and Bangladesh, the proposed Tipaimukh Dam is beginning to draw international attention. Large hydroelectric dams used to be considered one of the cheapest ways to generate power. Assessing them holistically and counting the externalities, they have now been proven to be disastrous – environmentally, culturally, economically and, not least, morally.

A concern is also growing over possible connections between large hydroelectric dams and earthquakes. On the one hand, earthquakes can break a dam, and on the other, by exerting enormous pressure on the earth's surface, large dams can also cause earthquakes, both with catastrophic consequences. Scientists of varied disciplines are beginning to investigate these connections. With aging large dams, more dams in the pipeline and a growing number of earthquakes around the world, their findings must be taken seriously.[28]

The battle against large-scale hydroelectric dams is still a battle against the tide. Fortunately, however, the opposition is growing and there is also a growing trend towards smaller, ecologically balanced, hydroelectric systems. Categories have evolved, such as 'medium' (10–100MW), 'small' (100KW–10MW) and even smaller 'microhydros' (also called 'minihydros').

An example of a microhydro is 'Lil Otto', which is 19cm round (excluding nozzle holder), 33cm high and weighs less than 4.5kg. It is made by Lil Otto Hydroworks in Hornbook, California. Placed in a stream or a spring, under optimal conditions, it has the capacity to generate electricity up to a maximum of 5amp/hr (120amp/day). It operates '24 hrs a day, no Sundays off, and the sun don't gotta shine!' It is a stand-alone system which, if needed, can easily be connected to another power source – PV or wind-turbine – to operate as part of a hybrid system.[29]

Artificial streams can also be used to generate electricity. In Coral World, a marine park located on the east end of St Thomas, US Virgin Islands, a microhydro taps the wastewater outfall from the marine display tanks to turn a turbine generator.[30]

Another innovative example is called the 'Tyson Turbine', named after its developer, the Australian farmer Warren Tyson. This turbine can both generate electricity and pump water powered by the energy of flowing water. It comes in 13 models to match the power need, budget and power source – which could be a stream, lake, creek, canal, river or any such source with the needed depth and speed of water flow. The Tyson Turbine does not require the use of dedicated civil structures to harness the energy of flowing water, which significantly accounts for the economy, portability and versatility of the system. At a site well-suited for its operation, compared to PV, the Tyson Turbine has been proven to be six times cheaper for electricity generation.[31]

So, it is possible to have a range of ecologically balanced hydroelectric systems. In China nearly 90,000 small hydros have been powering the country's rural industries.[32] India has given the technology a priority in its national renewable energy programme and already installed small hydros with a total capacity of 1341MW.[33] In the New England region of the US there were once many small hydros playing a vital role in the region's economy. During the fossil-*fooled* 'cheap fuel era' most of them were abandoned or they simply perished. Around the world, through careful reassessment and innovation, it is time to tap into the big potential of ecologically balanced hydros, including the smallest ones.

Solar collectors for hot water

In the most remote parts of the world one may be struck by the techniques with which people collect hot water. In rural Bangladesh I have seen large black clay containers filled with water and put out in the sun to work as hot water collectors. The same containers, when placed in shade, work as coolers. A rather simple and ingenious technological invention! It seems that there has always been a need for hot water and, in one form or another, there has always been a solar collector. Black-painted tin drums, too, have been used in the same way. So, there have been variations and there have been improvements.

In the US during the 1890s, Clarence M. Kemp, a Baltimore inventor and manufacturer, introduced his 'Climax Solar-Water Heater'. Combining the old principle of exposing bare metal tanks to the sun with the scientific principle of the hot box, it represented a breakthrough improvement for its increased ability to collect and retain heat. Known as the nation's first commercial solar heater, the Climax was available in eight sizes, from the most popular and smallest, a 32-gallon heater measuring 1.4m long, 1m wide and 0.3m deep, and sold for $25, to the largest, a 700-gallon heater, sold for $380. In each model, four cylindrical tanks made of heavy galvanized iron painted in dull black were placed next to each other inside a pine box insulated with felt paper and covered with a sheet of glass. The box was installed facing south on a sloped roof or placed on brackets at an angle to a wall, so the tanks lay horizontally one above the other. The tanks were filled with water, which was heated by the sun.[34]

In the US the use of solar hot water systems flourished, with a range of design and brand variations, until the industry became another casualty of the 'cheap fuel era'. In fact, households received special incentives to replace their solar hot water systems with fossil-fuel powered systems. The trend continued until the 1970s oil crisis. Then, supported by government tax rebates and other subsidies, the industry was revived. Even Sears sold solar hot water systems or provided a service warranty for systems installed through its subcontractors. Unfortunately, it was only a short-term boom, dissipating largely due to the propagandized 'end' of the energy crisis and the real end of government subsidies during the 1980s. This was an 'anti-renewable energy era' during which the solar hot water panels put up on the White House roof during the Carter administration were unceremoniously taken down during the succeeding Reagan administration. Sears stopped selling hot water systems. Fortunately, however, catalogues like Real Goods' *Alternative Energy Sourcebook* picked up where Sears left off. The catalogue has always been more than a listing of renewable energy products. Put out by some visionary activists, like John Schaeffer, it is a true activist's handbook which has grown in size, circulation and stature, and come to be known around the world by its current title, *Solar Living Sourcebook*.[35]

As the need for hot water has grown over the years, so have its costs and the percentage it consumes of the household energy budget. American households spend up to 25 per cent of their energy budget on heating water. Worldwide domestic, commercial and industrial needs continue to grow just as the need to find cost-advantageous alternatives to fossil-fuel powered options increases. In some industrially developing countries, wood burning is a means to heat water for small-scale industrial purposes, but the scarcity and cost of firewood continue to go up.

Modern solar hot water systems offer a variety of options and versatility in their uses. Built upon the long history of well-tested principles of the first generation of scientifically designed collectors, such as the Climax, they are more efficient than their predecessors. The two basic elements of a system are a solar collector and thermal storage. These can be integrated, as in a 'batch' heater – a water tank painted black and enclosed in a glazed, insulated box, and placed in the sun to heat the water inside the tank in one batch. Or, the elements can be placed apart from each other. Water can be circulated between the collector and storage utilizing passive or active systems. The passive system is a 'thermosyphon' system in which the water tank is installed above the collectors. Thermosiphoning is the natural movement of liquid (or air) as it is heated or cooled. Water inside the pipes running through the collector is heated, which then rises through piping to the tank to be displaced by cold water sinking to the collector – the lowest point in the circulatory system – to rise again after being heated. It is called 'passive' because it does not use any moving parts such as pumps, fans or other mechanical controls to distribute collected heat – as in an active system. The batch heater is one type of a passive system. One of the breakthrough developments in the technology is the blending of passive and active elements without the use of any pumps or controls. 'Copper Cricket' systems represent this development. They operate on a 'geyser pump' principle. Boiling methyl alcohol forces water up through the solar collector located as high as 36 feet above the tank and into circulation, similar to a coffee percolator.

Home-built batch heaters can cost as little as $300, depending on local material costs. A passive system adequate for a family of three people may cost around $1000. An active system adequate for about five may cost between $2000 and $4000. As long as the sun is shining, the system will operate; the percentage of the hot water needs being met, the 'pay back' rate, etc., depend on the system size, costs and climate. However, with free fuel costs and the 20-year or more life of an average system, solar collectors are proving to be increasingly attractive investments.

One of the advantages of having a technology around for many years is that it allows us to assess its worth based on actual use and results, not just lab tests and projections. Here is an example. Even in the New England climate – usually considered not so favourable for sunlight – a Sears Solar Hot Water System with a 100-gallon tank, bought for under $2300 full price 25 years ago, has been saving at least $250 annually for the Noce family living in a Boston neighbourhood. Not only has the system paid off the out-of-pocket expense for the family, which was about $500 after federal and state rebates, the full price of the system has been paid off. The system provides up to 100 per cent of the household's hot water needs during July and August, down to 25 per cent during November and December, for the family of four and guests.

The trend for improvements and innovation continues. Two things are clear: one, the need for hot water for domestic, commercial and industrial purposes – around the world – is growing; two, solar collectors – with a variety of designs and scales – are ready to play an increasingly vital role in the transition to the renewable energy path by satisfying that need.

Solar greenhouses

Worldwide there is a need to grow more food and we need to grow it safely. In fact, it is only by growing it safely – in the long-run – that we can grow more food. This is one of the most crucial lessons from recent practices in modern agriculture. In its quest for food abundance, resource-intensive agriculture, based on chemical pesticides and fertilizers, sprinkler irrigation, heavy machinery and monoculture, is actually proving to be its own antithesis. Beyond a few short-term benefits, such agricultural practices have been proven to be causes of – not solutions to – decreasing food production; inequities; nutritional, health and economic crises; and poisoning of the water, air and land. In order to grow more food safely and share it more equitably, we also need to learn how to grow food locally and self-reliantly. And we need to grow food in cold climates that normally interfere with agricultural food production.[36]

Solar greenhouses offer an ingenious and cost-effective option for growing plentiful and healthy food even during seasons which, literally, freeze agriculture. It is like extending the growing season, or adding a new growing season, to the annual calendar. Inside well-insulated, south-facing structures trapping heat from the sunlight, solar greenhouses transform that particular spot of land in the midst of a frozen, snow-covered landscape into a thriving, varied, lush, tropical growing field!

One of the most inspiring examples of such a greenhouse is Solviva Winter Garden on Martha's Vineyard, an island off the coast of Massachusetts. Designed and run by Anna Edey, the energy self-sufficient $372m^2$ solar greenhouse (which uses PV to power fans and lights) has grown a variety of organic food commercially and successfully since 1983. It has been productive throughout the year – during the coldest New England winters of well below 0°F, as well as the hottest summers. The many features of this ecosystem include design; glazing system; management of warm-blooded animals, such as chickens and angora rabbits, providing warmth, carbon dioxide, fertilizer, eggs, meat and angora fleece for high quality yarn; solar heating systems; harmonious insect management; multilevel growing; and composting.

Solviva has provided the understanding of how a greenhouse can operate naturally, non-violently and harmoniously with the Earth's biological processes – day after day, year after year. Lately Anna has taken that knowledge and skill to build smaller, home-scale solar greenhouses.[37]

Another example is the Badgersett Research Farm Greenhouse located in the extreme southeast of Minnesota, where summers are short and hot and winters are long and very cold. Badgersett is a family business, developed by former university researchers Philip and Mary Rutter and their sons, Brandon and Perry. The Rutters planted their first trees in the 1970s and, with that success, began large plantings in 1981. By the spring of 1993 they had about 6000 chestnuts and about 5000 hazels. The Rutters had grown more than 20,000 plants in the greenhouse that year, with even higher expectations for the future. The semi-earth-sheltered greenhouse was co-designed with Roald Gundersen, who also worked on Biosphere 2 in Arizona. The PV and wind turbine systems for the greenhouse were developed by Real Goods of Ukiah, California.[38]

Solar greenhouses can also be built as an extension of an existing house or building. The basic principle and structure are the same – a south-facing enclosed space with glazing to collect heat from the Sun. But the designs, sizes, uses and names vary: sunspaces, sun-rooms, atriums. Whether used as an extension of living space, or a flower, fruit or vegetable garden – or even a fish farm – a solar greenhouse can perform the dual function of a multipurpose physical space and a solar collector to heat a house.

Solar Cookers

> *The kitchen was situated in a beautiful vegetable garden. Every creeper, every tomato plant was itself an ornament. I found neither smoke, nor any chimney either, in the kitchen – it was clean and bright; the windows were decorated with flower garlands. There was no sign of coal or fire.*
>
> *'How do you cook?' I asked.*
>
> *'With solar heat,' she said, at the same time showing me the pipe, through which passed the concentrated sunlight and heat. And she cooked something then and there to show me the process.*

The quotation is from 'Sultana's Dream', a short story by Begum Rokeya (1880–1932), a pioneer of women's education and gender equality, writer and social reform-er, who was born in Pairabond in Rangpur (now in Bangladesh) and died in Calcutta (now Kolkata, India). The story, originally published in English in the *Indian Ladies Magazine*, Madras, India, 1905, is one of the earliest, if not the earliest, documented expressions of the scientific imagination behind solar cookers. The same story, a pungent satire on male-dominated society, which antedated by a decade the much better known feminist utopian novel, *Herland*, by Charlotte Perkins Gilman, is also a masterpiece of ecological and renewable energy literature that envisioned natural conservation, environmental protection and – get this – scientific advancements that included the use of solar electricity, solar heat collectors, rainwater harvesting and hydrogen-powered vehicles![39]

In terms of the practical use of sunlight for cooking, quite possibly the first cooked food was solar-cooked. Before our primitive ancestors learned how to build a fire, chances are that they were already cooking with the Sun – 'curing' or 'baking' or 'sun drying'. In that sense, behind the modern-day solar cookers we have a principle of applying direct sunlight to cooking that has been tested and perfected scientifically and technologically ever since one or more of our ancestors hit upon the idea of serving warm or dried fish, fruit, vegetables or meat![40]

However, because of our reliance on wood, dung, natural gas, oil, coal and electricity, cooking fuel shortages now are among the most critical and worsening problems in many parts of the world. And for sure – unless some fundamental changes are made – sooner or later all of us will face the problem, no matter what part of the world we live in. The taken-for-granted era of turning on the stove fuelled by gas being piped in from miles away is showing a downward curve.

Fortunately, solar cookers in a variety of types and designs are proving to be a promising solution to cooking – simply by sitting in the sun! Therefore, this technology also has the potential to play an increasingly significant role in solving other associated problems such as deforestation and pollution. A solar cooker can also serve many other purposes, such as pasteurizing water and milk; disinfecting medical equipment, bandages and septic wastes; preserving and drying food; and dyeing wool and cotton yarn. By understanding how a cooker works, it can be used for many additional purposes. Each type and design has its own set of advantages and disadvantages. Combining the use of different types would be one way to get the best of each.

One of the simplest solar cookers is called the 'Quicky CooKit', promoted by Solar Cookers International (SCI), Sacramento, a nonprofit organization based in Sacramento, CA. It can be built by following these simple, quick steps:

1 Cut a large (approx. 60cm square x 23cm high) top-open cardboard box diagonally so each half has two walls and a triangular bottom;
2 Tape a 23cm-wide extra strip of cardboard to the cut edge of the bottom;
3 Cover the inside of the half-box, including the strip, with aluminium foil. One box, therefore, makes two cookers;
4 Put food in a pot with a lid (black metal pots absorb and retain heat better than most other kinds of pots);
5 Place the pot on a pot stand made of wires, twigs, stones, etc. so the heated air can circulate under the pot, and put it inside a reusable high-temperature clear plastic bag to hold air all around and under the pot, closing the bag with a twister or string;
6 Place the pot on the bottom panel of the half-box facing the Sun. Left in the sunlight, accumulated heat – which can reach higher than the water boiling temperature of 210°F or 100°C – will be trapped inside the plastic bag, heating up the pot and slow cooking the food! The high-temperature plastic bag can be replaced by a clear glass bowl. Put the glass bowl over the pot like a dome, making sure that the bowl sits firmly on the base so there is no heat leak.

Another type is the 'solar box' cooker. Basically it is an insulated box trapping the heat from the sunlight entering through a transparent top cover. It can be built with a range of alternative materials, such as a cardboard box, wood, metal, brick, Plexiglass, etc. for the main body, and a regular window glass sheet for the transparent top cover. The inside of the box and the reflector over the glass opening of the box can be lined with a shiny material such as aluminium paint, aluminium sheet, aluminium foil or tin sheet to direct the sunlight onto the pot. If preferred, the inside of the box can be painted black, without causing a significant difference in temperature. As much as possible, use non-toxic materials. Crumpled newspapers, rice husks or coconut fibres can be used for insulation. The box can be sized to accommodate several pots, therefore cooking several dishes at the same time. On a good sunny and warm day, sitting south-facing, the temperature inside a solar box cooker can go as high as 350°F – well above the boiling temperature of 210°F and capable of cooking a wide variety of dishes. It is also called a 'Solar Oven' or a 'Hot Box'.

Solar cookers cook food nutritionally and, with some types, without the need for anyone to attend to them. They really do the cooking for you! Left inside the solar box cooker, the food will stay warm for several hours after the Sun has set. Solar Cookers International (SCI) is devoted to promoting simple, effective and inexpensive solar cookers around the world (see Additional Resources for additional information on SCI).

There is also the 'Concentrator' type cooker. In the shape of a parabolic dish or an umbrella, its shiny interior facing the Sun reflects the heat as a concentrated beam to a particular spot where the cooking pot is hung. It requires more cautious handling and adjusting to the moving Sun. One of the advantages of this type over the box-type cooker is that it can be an almost instant high-temperature heat source – as high as 600°F, pretty much like lighting a match. A concentrator type solar cooker can be made of aluminium or other metals, or with a wooden frame with a shiny interior lining. Sizes can vary, from home to institutional scale. Costs will vary accordingly. Some concentrators have been designed so they can be folded like an umbrella for portability and storage. Beth and Dan Halacy of Lakewood, Colorado, have owned one for over 20 years and call it 'Umbroiler'.[41]

Beyond the initial purchase or material costs, a solar cooker will cook for free – with sunlight as its fuel. How well it cooks depends on the type of cooker and on the duration, angle and quality of sunlight. As long as there is sunlight, there will be some warming effect. However, the outdoor temperature, the seasonal angle of the Sun and geographical location are all important variables. In New England, the period between mid-spring and mid-fall is the best season for solar cooking – although it has been possible to bake bread in a very well-insulated solar box cooker on a sunny day during the snow-covered winter month of December. In Armenia, treats of pre-dried fish have been prepared in a solar box cooker during the middle of winter! In some parts of the world, solar cookers are usable almost year-round. This flexibility is especially useful in the dry regions with scarcity of wood and other cooking fuels.

Most solar cookers need direct sunlight to cook, therefore, they cannot be used all the time. Whether on a daily or seasonal basis, developing the habit of solar cooking when sunlight is available can be a challenging one. The habit can be as simple as putting out the cooker in the sun before going to work and bringing it in when you get home. Still, requiring a change in the habit of cooking on a conventional indoor stove is a challenge. But by educating ourselves about the importance of solar cookers and using them as much as possible we can cut down on using up nonrenewables and save them for the times when sunlight is not available. Also, whenever and wherever possible, the use of solar cookers requiring direct sunlight can be combined with other renewable energy technologies for cooking, such as cookers fuelled by biogas and electric stoves powered by electricity generated through renewable sources. That is a growing and innovative option.

Revolutionary breakthroughs are heating up solar cookers too, day or night! The Tulsi-Hybrid Solar Cooking Oven cooks no matter what the weather is. Developed in India, priced between $250 and $300, it was introduced in the US through various outlets in 2006. Designed on the principle of a solar box cooker, with the attractive look of a Samsonite suitcase, in direct sunlight it generates temperatures up to 400°F. When there is no sun or at night, it can be plugged into a standard 120VAC outlet to cook using 75 per cent less energy than a conventional oven.[42]

Tulsi – and I am sure more of its kin will follow – can be made 100 per cent renewable-energy powered if the electricity comes from a renewable source. In a home with a utility-connected renewable energy system already inverted to 120VAC current, it will require no adjustment. To cook from a stand-alone system with proper capacity powered by a renewable source, usually 12VDC, the heating element can be replaced to match the source. Or, as we do with our Tulsi and a stand-alone PV system installed on a south-facing window sill of our fifth-floor apartment in Boston, an inverter can be used to invert the 12VDC power generated by the PV system to match the 120VAC of the oven. Rain or shine, day or night, around the year, dinner will be ready – and solar cooked too!

Biogas plants

As plants and animal wastes undergo bacterial decomposition in the absence of oxygen, called the 'anaerobic condition', they produce methane, an odourless, colourless, flammable gas. In the presence of oxygen, the 'aerobic condition', they produce compost. This naturally occurring process, as old as the cycle of life itself, is the science behind the renewable energy technology called the 'methane generators', 'biodigesters', or 'biogas plants'. All kinds of animal and human wastes – such as cow dung, chicken droppings and human excrement, vegetable wastes and agricultural residues can be used to generate biogas. The system design and scale can be flexible to meet the needs and resources of a family, a group of families or an entire community. Also, different construction materials can be used.

The largest number of biogas plants is in China. More than 5.5 million household-scale plants are in use in the country.[43] In southwest China's Guangxi Zhuang Autonomous Region, more than 700,000 biogas plants have been installed – one in every ten farmhouses.[44] China has also constructed some industrial-scale biogas plants.

The second largest user is India. More than 3.1 million biogas plants are used in Indian villages.[45] In other developing countries, especially due to a growing scarcity and rising costs of fuel wood, kerosene and natural gas, biogas plants are gaining popularity. Among such countries are Nepal, Indonesia, Bangladesh, Sri Lanka, Thailand, Vietnam, the Philippines, Cambodia, Colombia, Morocco, Ethiopia, Benin, Burundi and Tanzania.

It takes three cows worth of cow dung, or equivalent fuel, to generate enough biogas for a family of four. That limits the option of ownership of a plant to a few relatively well-off families. In such situations, co-operative ownership of larger plants for several families can be an answer. However, the technology is becoming more efficient. There is some indication that the wastes needed to generate the same amount of biogas can be reduced significantly in the future, as much as by one-third, which will make the technology significantly more attractive and affordable. Actually, Dr Anand Karve of the Appropriate Rural Technology Institute (ARTI) in Pune, India, has developed a compact biogas plant which can generate enough cooking gas for all the meals of a nuclear family from a daily supply of only 2kg of vegetable scraps. Also, it is designed so that the drum-like biogas plant can be placed either on the ground or a roof – a revolutionary boon to urban dwellers of multi-storey buildings. Especially since receiving the prestigious Ashden Award for Sustainable Energy in 2006, the ARTI biogas plant is drawing significant international attention.[46]

Material and labour costs can vary but, generally, biogas plants are highly cost-effective. In Bangladesh, a family-size biogas plant made from locally available materials like bricks, cement and plastic pipes and valves, with an expected 30-year life and no planned obsolescence, costs between Taka 12,500–15,000 ($250–300). At the rate of Taka 300 per month, which is about what a rural family spends on firewood per month, the cost is recoverable within four years. Microcredit stretches the total amount and the length of recovery, but it also enables a family to own a plant without needing to pay the full price upfront and pay off the loan from the money saved from firewood expense. A government subsidy of Taka 5000 per plant has made it more affordable for families.

Every technology has its own set of challenges and limitations, but the benefits of biogas outweigh its costs by far. In addition to fuel for cooking, biogas plants provide byproducts such as very high quality organic fertilizer and fish feed. They preserve topsoil, improve soil quality, and save trees and plants from being cut down for firewood. They recycle wastes which would otherwise cause pollution and other health hazards. Biogas can also power gas lamps and generators to produce electricity and heat.

Furthermore, not only are biogas plants a most promising technology for agricultural settings around the world, they also have the revolutionary potential for reviving a sustainable agricultural lifestyle where such a lifestyle has been lost or threatened. Thus, this technology may become an important technology of the future, not only for developing countries, but also for developed countries. Chemical agriculture, fuelled by nonrenewable energy, is turning land into barren and toxic wastelands, where the cycle of life needs to be urgently revived. Results from biogas projects of various designs and scales – for homes, communities and industries – generating cooking fuel, heat and electricity and organic fertilizer as a byproduct, in countries such as Austria, Belgium, Canada, Denmark, Germany, the Netherlands, the UK and the US, suggest that farmland revival – a sustainable agricultural revolution – is indeed possible.[47]

A Brighter Future – Sprouting Now!

So, there we have it, in brief but demystified, a set of renewable energy technologies – photovoltaics, wind turbines, hydroelectric generators, solar collectors for hot water, solar greenhouses, solar cookers and biogas plants – which have proven their merit. Each of these offers a wide range of application choices, from a small and personal scale to the thousands of times larger community scale. And these are available, or can be made available, around the world right now. Nothing better describes – or practically demonstrates – what 'power to the people' means.

Without question, to make a transition to the renewable energy path we will need more. There are more options and, as innovations and breakthroughs continue, there will be many more to come. Some emerging options, in the sense of their need to undergo some breakthroughs for widespread use, lead the way.

Hydrogen fuel cells

Probably the most talked about emerging option is hydrogen fuel cell technology. Water is split into hydrogen and oxygen, then recombined inside a fuel cell to produce electricity.[48] The fuel cell stores the electricity, which can be used for a wide variety of purposes, designs and scales – from running cars and boats to powering homes and factories.

The promise of hydrogen fuel cell technology is revolutionary. It is a more efficient and cleaner alternative to the combustion of gasoline and other fossil fuels. It uses water, a more abundant and sustainable source. It is simpler and quieter than conventional methods of electricity generation because it has fewer moving parts, and is thereby also more durable and less prone to wear and tear from operation. And it is emerging as an ideal solution to the problem of storing large quantities of energy from renewable sources. The Sun does not shine around the clock and wind does not blow all the time. Their daily intermittence, along with seasonal variations, therefore presents a challenge when it comes to storing energy so that it can be delivered consistently on demand. The revolutionary potential of the high capacity fuel cells to store large quantities of energy in a compact way is emerging as an answer. The exhausted Genie, bottled in the conventional energy sources, is regaining life in renewable sources and finding a new home in fuel cells! But because of the high upfront capital cost, it is still far too expensive to be practical on a large scale. And for the electricity needed for splitting water, or electrolysis, it depends on nonrenewables – thereby replacing one problem with another. Over time, however, all that can change. The production scale, technology and efficiency can improve, lowering the cost, and the electricity needed for electrolysis can be derived from renewable sources such as PV or wind turbines. An infrastructure for mainstream utilization of this technology is evolving. Pilot projects – such as cars and buses using fuel cells – and new research findings around the world are lighting the way.

Sensing the huge potential, university and corporate labs are hotbeds of fuel cell research. One of the recent breakthroughs took place at an MIT lab. Professor Daniel Nocera, principal investigator of the Solar Revolution Project and co-director of Eni-MIT Solar Frontier Center, and his team have developed a simple, inexpensive and highly efficient method to split water to produce and store electricity inside a fuel cell. There is more work to be done before its widespread use. 'This is just the beginning,' says Nocera. He hopes that within ten years people will be able to power their homes and cars with fuel cells, using photovoltaics to perform the electrolysis.[49] Nocera also envisions, with rooftop PV modules fuelling the cells, grids will no longer be necessary, eliminating one of the most cumbersome, expensive and hazardous components from the energy distribution system. For the developing world, where up to 70 per cent of the area is beyond the reach of the existing power lines, and in which the vast majority of the population lives, the advantage will be revolutionary.[50]

Biofuels

There's been quite a bit of talk about biofuels. They are derived from biological resources such as plants and agricultural products, residues and wastes.[51] The most well-known is corn-based ethanol, which can be blended with gasoline, as a transportation fuel. Especially since US legislation mandated in 2007 the annual production of 36 billion gallons of biofuels by 2022, investors and big Midwest farmers (supported by huge subsidies and powerful lobbyists) have jumped on to the profit-driven and politically motivated biofuels bandwagon in a way similar to the discovery of an oil field. The European Union (EU) too has a plan to fuel 10 per cent of the cars and trucks in the EU with biofuels by 2020. Hyped up almost as a panacea for transportation fuel need, biofuels have just as quickly become a highly controversial topic. Rapid diversion of corn, soybean, sugar cane, canola, maize and other agricultural items into ethanol production has shot up the prices of food made out of those items. So, it has become a 'food vs. fuel' issue. The impact is worldwide and those who suffer most are the poor. With a growing need for food – which people need more than cars – the impact is crucial. It is a moral issue, as well. Protests have erupted around the world. Demand has also pushed up the prices of a host of other nonfood products which use these items in their production mix. It has caused rainforest destruction – from Brazil to Malaysia – and clearing of forests and grasslands, converting them into biofuel crop fields. Consider also the destruction of natural biodiversity. Attempts to promote biofuels as an eco-friendly, low-carbon energy resource have found their critics. Clearing of forests also releases carbon, as does the intensive use of fertilizer to grow corn and other agricultural items for biofuels, biofuels production processes and transporting the biofuels. Everything added up, the net gain against conventional gasoline may be marginal, even negative.

So, is there hope beyond the hype for biofuels? Advocates claim yes. They point to the use of agricultural wastes – like sugar cane bagasse biomass and wheat straw – and residues and feedstocks from nonagricultural lands, like algae. They propose harvesting nonfood plants, like native switchgrass, for biofuels production. Nonfood biofuels have a new name, 'Biofuels 2.O'. Companies around the world see a green future for algae production. Glen Kertz, CEO of El Paso based Valcent Products, which is working on high yield, closed loop, algae production, says, 'If we took one-tenth of the state of New Mexico and converted it to algae production, we could meet all the energy demands for the entire United States.'[52]

All this leads to confirming that there is an acknowledgement in the scientific community – and green business community – that, with careful assessments of priorities, standards and guidelines, sustainable harvesting, scale of production, and more efficient production processes and distribution, biofuels can be certified as an eco-friendly fuel source in our energy mix, resolving the biofuels dilemma. So, we should be careful 'not to throw the baby out with the bathwater'. At the same time, because there are conditions which must be met before biofuels can be certified as a resource to be promoted on a large or appropriate scale, we must also not put the 'cart before the horse', which the US and EU biofuels policies have essentially done. These policies must be reconsidered. Then, after the horse has been put before the cart, let the journey continue!

Geothermal

Geo (Earth) and thermal (heat) – combine the two with the help of technology and we have a geothermal system. Here are some basics. As the Earth absorbs heat from sunlight, it is stored deep in the rocks and groundwater. It happens on a daily basis, as well as due to the temperature built up over time at the Earth's core, 6,400km below the surface, with temperatures as high as 9000°F. The core releases the temperature outwardy, causing rocks to melt and surface as lava flows, volcanoes erupt, fuelling geysers and hot springs, and heating up rocks and ground water. A geothermal system, powered electrically, injects water down through a pipe designed as a loop to the heated zone of rocks and ground water – the 'heat source' – then pumps up the heated water to the surface. It can extract and distribute the heat evenly at a steady temperature. The temperature can also be raised to produce steam to power a turbine to generate electricity. In a closed-loop system, the water is re-injected into the Earth to repeat the cyclical process. The system can also reverse the process for cooling. It can extract heat from the air, like in summer, and send it deep inside the Earth. The Earth now serves as a 'heat sink'. For its reliance on the Earth's renewable heat source, the geothermal system is considered a renewable energy technology. Capable of heating and cooling – both of space and water – and generating electricity, geothermal technology is impressive and its popularity is growing among private home and institutional building owners. A geothermal industry is poised to flourish. Estimating from various sources, starting with the first recorded geothermal system installed in Italy in 1904, the technology is being used in at least 50 countries. Iceland, endowed with its geography of hundreds of volcanoes, geysers and hot springs, which make the technology most appropriate for the island, leads the way.

The potential of geothermal technology is huge, no doubt. But, like with any technology, we must proceed with caution, under carefully crafted guidelines. The technology and how it works may seem simple, but the ecology of the Earth is more complex and interdependent than viewing the drilling and extracting heat as isolated factors. Geological makeup, permissible depth for drilling at a specific location, underground ecological impact of drilling, the rate of heat and steam extraction, water usage and an analysis of an unintended release of gases from the Earth's interior – all these and more factors must be seriously considered. The critical environmental consequences of reckless exploration and exploitation of nonrenewable resources warn us that human beings, especially with our industrialized and compartmentalized consciousness, understand very little of the Earth's complex ecology and equilibrium. When such holistic wisdom, even a scientific voice, has spoken, it has often been undermined, if not deliberately suppressed, under the aggressive drive for extraction and exploitation, resulting in our current crisis. 'Drill, baby, drill!' and 'Consume, consume, consume!' still represent a pervasive attitude toward the Earth and its resources, backed by a lot of powerful vested interests. If that attitude can be changed or, at least, can be effectively controlled by guidelines, geothermal will certainly prove to be a significant and sustainable energy source.

Hydrogen fuel cells, biofuels and geothermal – these are just three of the emerging options. While allowing them the time they need to develop through addressing the current concerns and limitations, we can utilize to the fullest the technologies we have at hand to set the transition in a fuller motion and build an infrastructure that will support and facilitate the emerging options. That infrastructure will have to be built on diversity and balance. Diversity of the growing range of renewable energy technologies, designs, scales, and of appropriateness will have to match the diversity of needs, capacities and locations around the world. And there will have to be a balance between the utilization of a technology and sustaining solutions so that a short-term solution does not lead to a new and bigger long-term problem. The balance must be attained from a global perspective, so that a solution or benefit in one place does not result in a problem or undue expense elsewhere. In such equitable diversity and balance lies the real scope of renewable energy technologies. The renewable energy technologies available right now are seeds – firmly rooted and well-tested – but compared to their revolutionary potential, they are barely sprouting. We must let them flourish!

3

'A Picture is Worth a Thousand Words'

'I'll believe it when I see it.' Who hasn't, at one time or another, heard that phrase in response to an attempt to convince someone about something unfamiliar to that person, especially if that 'something' happens to be something out of the ordinary? That is a response I have received many times over the years when I talked about renewable energy technologies, whether to a person or a group – even about some simple applications like cooking with sunlight or lighting up a couple of rooms by putting up a small solar panel on the roof. The simplicity itself has sometimes been a reason for scepticism. And among the sceptics have been not only novices, but also policymakers and conventional energy experts. Fortunately, as actual implementation of some of the technologies has sprouted around the world, and information about them has become more accessible – prompted by a growing awareness of the energy crisis – the spirit of the comment has changed from disbelief and scepticism to receptivity and curiosity.

But the challenge remains if we want to expedite the transition to a renewable energy path. There is a real need to show, not just tell about, the incredibly diverse ways that renewable energy technologies have not only evolved, but actually been implemented. There is a need to demonstrate the diversity of the renewable energy technologies – among which are **photovoltaics, wind turbines, hydroelectric generators, solar collectors for hot water, solar greenhouses, solar cookers and biogas plants;** the diversity of design and scale of each technology; the diversity of the purpose of their applications; and the diversity of the locations around the world where they have been implemented. Together they create a scenario that is simply beyond words. It is dynamic, inspiring, empowering and, indeed, revolutionary.

Here's a glimpse. Let each picture speak a thousand words!

Photovoltaics

Figure 3.1 *Stand-alone PV-powered, passive solar home of solar pioneer Paul Jeffrey Fowler, constructed in the 1980s, in rural Worthington, western Massachusetts*

Source: Sajed Kamal

Figure 3.2 *Rural home powered by a stand-alone PV system in Bangladesh. 100 per cent renewable-powered, it also has a Hot Box and a biogas plant for cooking*

Source: Sajed Kamal

Figure 3.3 *PV-powered medical refrigeration unit in a Kenyan desert*
Source: Mobil Solar, courtesy of Massachusetts Photovoltaic Center

Figure 3.4 *PV-powered desalination centre in Jeddah, Saudi Arabia*
Source: Mobil Solar, courtesy of Massachusetts Photovoltaic Center

Figure 3.5 *In New York City's Times Square, this 48-storey skyscraper has a 'PV skin' on all four sides which replaces the traditional glass wall. The building also incorporates other features like fuel cells and daylighting*

Source: Sajed Kamal

Figure 3.6 *Solar-powered health clinic in rural Cuba*

Source: Cubasolar

Figure 3.7 *PV-powered automobile charging station at the University of South Florida*
Source: Prof. Elias Stefanakos, USF

Figure 3.8 *Boston Nature Center, an urban education centre managed by the Massachusetts Audubon Society, includes grid-connected PV roof shingles, pole-mounted PV arrays, PV-powered street lights, a solar hot water system, a geothermal heat pump, daylighting and energy-efficient features*
Source: Sajed Kamal

Figure 3.9 *A lighted bulb, a toy boat and a cooling safari hat – all powered by PV*

Source: Sajed Kamal

Figure 3.10 *When this 100 kWp PV system was built on top of an existing sound barrier 20 years ago, it was the first PV project along a motorway worldwide and the largest PV plant in Switzerland. Situated along the A13 motorway in the Swiss Alps, the project combines noise reduction with energy production requiring no additional space.*

Source: ©TNC Consulting AG Switzerland, www.tnc.ch

Figure 3.11 *The Michael E. Capuano Early Childhood Center in Somerville, Massachusetts, has a 35kW PV system mounted flat on the roof*

Source: Sajed Kamal

Figure 3.12 *Boston Mayor Thomas Menino (third from right) inaugurating a 2.4kW PV system at the Boston Arts Academy, a public high school, in the Fenway on 6 April 2006. A Solar Fenway project*

Source: Rosie Kamal

Figure 3.13 *A 100kW PV system and a 10kW wind turbine, both utility-interactive, at Beverly High School in Massachusetts. The site meets more than 9 per cent of the school's annual electricity needs, saves $10,000–20,000 yearly for the city of Beverly and is used as an educational tool for students and teachers around the world. A Solar Now project*

Source: Sajed Kamal

Figure 3.14 *Water delivery system powered by PV modules supplied by Soluz Dominicana, S.A. to serve a rural community in western Dominican Republic.*

Source: Soluz, Inc., www.SoluzUSA.com

Figure 3.15 *A rural home with a stand-alone PV system in El Salvador*

Source: Sajed Kamal

Figure 3.16 *Two of the 30 PV-powered homes in Gardner, MA. 'The first solar electric neighbourhood in the world', each with a 2kW grid-connected system. Gardner also has PV systems, up to 5kW each, on its Town Hall, library, community college, a furniture outlet and a Burger King*

Source: Sajed Kamal

Figure 3.17 *A solar-powered home in Worthing, West Sussex, UK*
Source: Sajed Kamal

Figure 3.18 *Catamaran 'sun21', a 14.6m boat completely powered by a 65m²*
photovoltaic array, left Seville, Spain, in November 2006 with five Swiss crew
members on an 11,250km journey to arrive in New York on May 8, 2007,
demonstrating the great potential of renewable energy. The boat is also noiseless
and does not pollute air or water

Source: www.transatlantic21.org

Figure 3.19 *A wall-mounted PV system at Fenway Views Condominium-Affordable Housing Building of the Fenway Community Development Corporation (FCDC) in Boston, MA. A Solar Fenway project, 2007*

Source: Sajed Kamal

Figure 3.20 *A mobile street sign powered by a stand-alone system in Brookline, MA. PV has revolutionized portable street and highway signs used by transportation departments, police and construction companies across the US*

Source: Sajed Kamal

Figure 3.21 *The house of Dr Gert D. Rust in Koenigswinter, in the western part of Germany near Bonn: 12kW, grid-connected*

Source: Imke Buchholz

Figure 3.22 *1.3MW Solar park in Sinzheim, Germany. The modules are angled so that the area underneath can be used for grazing, allowing multiple benefits. As the sun moves, the entire area underneath receives sunlight*

Source: First Solar Inc.

Figure 3.23 *Solar-powered condominium building in Haar,*
a village near Munich, Germany

Source: Martin Wittmann

Figure 3.24 *The Lewis Environmental Studies Center, Oberlin College, Ohio. A 59kW,*
325m² PV array and a 100kW, 818m² PV array power the academic facility and
parking pavilion. A carbon-neutral, net energy exporting facility.
Construction period, including the BIPV system: 1998–2000

Source: Prof. John Petersen, Oberlin College

Figure 3.25 *Albe-res (renewable energy sources tree) in the City of Vasto, Chieti province, Abruzzo region, in the south of Italy. It is a 1.4 kW PV system, integrating public lighting with LED bulbs, charging point for small devices (like phones, laptops, electric bikes, etc), wi-fi internet (thanks to an antenna integrated on the top of the tree), and video security (the tree is also able to carry in its leaves a CCTV camera, when requested by the public administration).*

Source: Albe-res Srl, abertello@olicar.it

Figure 3.26 *Pal Town Solar City (Pal Town Jozai no Mori), Ota City, Gunma Prefecture, Japan. A 4-kilowatt system each for 550 homes has been installed, fully funded by the government, as part of a nation-wide effort to promote solar and other new energy technologies in Japan.*

Source: Ota City, http://www.nedo.go.jp/enetai/sinene_hyakusen.html

Figure 3.27 *Solar-powered Porter Square Shopping Center, Cambridge, MA*
Source: Sajed Kamal

Figure 3.28 *BedZED—the Beddington Zero (fossil) Energy Development—
the UK's best known eco-village in Sutton Borough, in South London.*
Source: Tom Chance, BioRegional Development Group, www.bioregional.com

Figure 3.29 *The Vandermark family in front of their home with a 7kW solar slate-roofing system in Cambridge, Massachusetts*

Source: Sajed Kamal

Figure 3.30 *Rohan and Juliana in front of the Solar Peace Fountain at the Somerville Community Growing Center in Massachusetts. Committed to environmental education and cultural festivities, the centre also has a PV-powered garden shed, a water pump and a solar greenhouse. It hosts the annual 'Boston Solar Day', an educational event celebrating the Sun (organized by the Boston Area Solar Energy Association, www.basea.org)*

Source: Sajed Kamal

Wind turbines

Figure 3.31 *A charging station in the village of La Barra de Rio Grande, Nicaragua. The tower on the left has an array of 12 solar panels and has proven to be a good hybrid system with the wind generators. The building also serves as a hurricane shelter and saved 75 lives in November 2009, when this tiny village was razed by Hurricane Ida*

Source: Peter Coleman/Peace and Hope Trust

Figure 3.32 *A wind farm in Lower Saxony, Germany, growing food and energy simultaneously.*

Source: Philip May, en.wikipedia.org/wiki/File:Windpark-Wind-Farm.jpg

Figure 3.33 *In Boston Harbour, Hull Municipal Light's Hull Wind 1, a 660kW wind turbine generates enough electricity to power 250 homes*

Source: Sajed Kamal

Figure 3.34 *Nysted Wind Farm, an offshore wind park off of Denmark.*

Source: Cape Wind, www.capewind.org

Figure 3.35 *A 300-watt AIR 303 wind turbine provides electricity for some of the rooms in this cyclone shelter in Pekua, in the coastal area of Bangladesh. A BRAC project*

Source: Sajed Kamal

Figure 3.36 *Tomamae wind farm in Hokkaido (North Pacific coast), Japan, is the largest wind farm in Japan. The first large-scale commercial wind power generation in Japan started here in 1999 when Tomen Corporation built 20 1000kW wind turbines. Since then, the Electric Power Development Company Ltd has added 30,000kW*

Source: Curt Davis

Figure 3.37 *GUS Vertical Axis Wind Turbine: overall height 2.9m, blade diameter 1m, weight 363kg, rated output at 60kmh, 11,000kWh/year. Accompanying PV arrays, 10kW*

Source: Cross Sound Builders, Corp.

Figure 3.38 *Wind-PV Hybrid System at IBEW 103 Training Center, off the Southeast Expressway, Boston: 100kW wind turbine, 45m, 5.4kW PV*

Source: Sajed Kamal

Figure 3.39 *Bahrain World Trade Center is the first in the world to install wind turbines placed in the wind tunnel created by two adjacent high-rise buildings.*

Source: Bahrain World Trade Center, www.bahrainwtc.com

Figure 3.40 *Wind turbines at Boston's Logan Airport. 20 AeroVironment roof-mounted turbines, 1.8m high, generate 100,000kW hours annually*

Source: Massachusetts Port Authority

Hydroelectric systems

Figure 3.41 *An ecologically designed 'pico-hydro' system in a remote community on the slopes of Mount Kenya*

Source: Martin Wright/The Ashden Awards ©2010

Figure 3.42 *A microhydro system in Guatemala. A project of Appropriate Infrastructure Development Group (AIDG)*

Source: AIDG

Solar collectors for hot water

Figure 3.43 *A Sears Solar Hot Water System at the Noce family home in a Boston neighbourhood. The $2300 system with a 100 gallon tank, installed in 1983, has been saving at least $250 annually*

Source: Sajed Kamal

Figure 3.44 *A solar hot water system near the Parthenon in Athens, Greece*

Source: Clayton Mendonça

Figure 3.45 *Solar hot water system at Fenway Park in Boston. Installed in 2008, the system provides about one-third of the park's annual hot water needs*

Source: Sajed Kamal

Figure 3.46 *Solar hot water systems on a hotel near Beirut, Lebanon*

Source: Virginio Mendonça

Figure 3.47 *Solar hot water systems at the Indian Institute of Management Calcutta (IIMC), Joka, West Bengal, India*

Source: Ashok Kamal

Figure 3.48 *A solar hot water system in Madagascar, an island nation in the Indian Ocean off the southeastern coast of Africa.*

Source: Edward C. Joyce

Solar greenhouses

Figure 3.49 *More than 600 robust solar greenhouses have been built in Ladakh, in the Himalayas, India. The greenhouses enable villagers to grow vegetables throughout the year for personal consumption and income generation—even when temperatures drop to -25°C. (L) Solar greenhouse, Ladakh. (C) Harvesting fresh vegetables even in winter. (R) Greenhouse owners have surplus vegetables to sell.*

Source: The Ashden Awards for Sustainable Energy, www.ashdenawards.org

Figure 3.50 *A 10kW grid-connected PV system and a solar greenhouse at Natick Community Organic Farm, Natick, Massachusetts*

Source: Sajed Kamal

Figure 3.51 *Winter growth inside the Natick Community Organic Farm*
Source: Sajed Kamal

Figure 3.52 *The permaculture homestead of Kyung Kim and Doug Clayton
in Jaffrey, New Hampshire. The house is designed as a super-insulated passive solar
collector providing heat via sunspaces and with other integrated systems like rainwater
collection/cistern storage, a root cellar and a home-designed and -built composting toilet.
Doug is currently researching the integration of home-scale biochar systems
(www.biochar-international.org)*
Source: Doug Clayton

Solar cookers

Figure 3.53 *A dish-type solar concentrator cooker in Manang, Nepal, a remote mountain village located 3500 metres above sea level*

Source: Gaurav Kotwal

Figure 3.54 *A Solar Quicky CooKit: a solar cooker made from one-half of a television box, lined with aluminium foil and with the pot placed inside a high temperature oven bag, can generate about 300°F! The cooker is also foldable*

Source: Rosie Kamal

Figure 3.55 *Solar cooker demonstration by Mujeres Solare de Totogalpa (Solar Women of Totogalpa). A joint project with Grupo Fenix, Nicaragua*

Source: Susan Kinne/Grupo Fenix

Figure 3.56 *A solar box cooker and a 'Tulsi-Hybrid Solar Cooking Oven'. Tulsi cooks day and night, rain or shine. It heats up to 400°F under direct sunlight. It can also be powered by plugging it into a standard 120VAC outlet, or, with an inverter, by connecting it to a stand-alone renewable energy system*

Source: Alex Roth

Biogas plants

Figure 3.57 *Cooking and lighting with biogas in Bangladesh. The biodigester, fed by cow dung and located underground outside the kitchen, generates cooking gas to be piped into the kitchen stoves and lamps*

Source: Rosie Kamal

Figure 3.58 *Biogas plant developed by Dr Anand Karve at the Appropriate Rural Technology Institute (ARTI), Pune, India. The compact, above ground, plant – placed on the ground or a rooftop – can generate enough biogas to cook all the meals for a nuclear family from a daily supply of only 2kg of vegetable scraps*

Source: ARTI

Figure 3.59 *A biogas plant at CIPRES, part of the Universidad Nacional de Ingenieria (National Engineering University), Managua, Nicaragua. It is a simple and fully operational biodigester that is connected to a latrine. The gas is used for cooking and the end waste material is compost used in gardens*

Source: Curt Davis

Figure 3.60 *Highmark biogas plant in Alberta, Canada. Cow manure is fed into the domelike anaerobic digester. Methane and carbon dioxide are then transferred to a power plant linked to the power grid. The plant produces high-grade fertilizer and generates 1MW of electricity.*

Source: Electrigaz Technologies Inc., www.electrigaz.com

4

From the Collapsing Economy to a Sustainable Economy: The Real Economic Advantages of Renewable Energy Technologies

---

There is no economic problem and, in a sense, there never has been. But there is a moral problem.   E. F. Schumacher[1]

Civilization as we know it will come to an end sometime in this century unless we can find a way to live without fossil fuels.   David Goodstein[2]

The first question I am often asked when I speak anywhere about renewable energy is about its cost. I welcome this question. Not only among renewable energy critics and sceptics, but also among its advocates, I find this to be one of the most misunderstood aspects of renewable energy. Actually, one of the crucial advantages of renewable energy lies with its cost.

In response, with appreciation, I ask if there are people in the audience who would agree with me on the general principle that we really should not be using something if we cannot afford to pay for it. Usually there is an affirmative nodding from everyone. Then I propose, if we are really serious about what we have just agreed on, we should act on it by turning off the lights, heaters, fans, air conditioners and whatever else we are using right now, right where we are sitting, say, in the middle of Boston. If that is too drastic, I suggest, we could at least turn off some and turn down some to minimize their use. Because, in a real sense, we simply cannot afford them. And if we insist on using them, we must ask, at what real cost? And who pays the price? Through a few moments of a mixed reaction of surprise, disbelief and puzzlement, this is usually an engaging prelude to addressing the question of renewable energy costs.

The total cost of energy

So, what is the cost? Dictionaries define it as the amount paid or to be paid for a purchase. Mostly we interpret it in terms of the money value directly attached to the purchased item, whether we pay outright from our pockets, with credit cards, or through instalments. But is there more to the cost beyond the directly stated amount? On the question of energy, is there more to the cost of electricity or heat beyond what's in the utility bill or on the gasoline pump receipt?

Climate change, acid rain, smog, oil spills, nuclear wastes and accidents, and the destruction of wildlife, natural resources and properties, which have now become proven and predictable consequences of utilizing nonrenewables, also incur economic costs which eventually fall on consumers. Such costs have been building up in our economy over the years. Some of these have been documented, even if ignored. In the late 1980s, one estimate of such environmental costs of energy consumption each year in the US was put between $125 to $150 billion. These costs included $7.5 billion from lung-related diseases from fossil fuel and more than $30 billion from radioactive waste.[3] 'The Rising Cost of Global Warming', a US Public Interest Research Group (US PIRG) study, estimated that the extreme weather-related disasters occurring during the eight-month period between 1 January and 31 August 1998 cost the US economy $14.4 billion in economic losses, $5.3 billion in insured losses and $1.3 billion in governmental disaster assistance.[4] A report by the United Nations Environment Programme (UNEP), released in February 2001, estimated the global cost of climate change to be $304.2 billion a year.[5] Another report, released in November 2006 by the World Meteorological Organization (WMO), the UN weather agency, said greenhouse gases reached a record high during 2005. The report coincided with the release of a British government report of a study led by Sir Nicholas Stern, former chief economist of the World Bank, which said global warming could cost the world economies up to 20 per cent of their gross national product.

Was Hurricane Katrina caused by global warming? This is no longer a headline, but both discussions and debates are still going on about a definite link between the two. The link deserves credence because Katrina matches so many of the consequences of global warming that scientists have been warning us about. The World Meteorological Organization and the National Aeronautics Space Administration (NASA) announced 2005 to be one of the warmest years on record and the US National Oceanographic and Atmospheric Administration labelled 2005 to be the costliest and deadliest year for hurricanes. These records lend much support to the opening lines of the 30 August 2005 *Boston Globe* op-ed piece, 'Katrina's Real Name', by Ross Gelbspan, Pulitzer Prize-winning journalist and author of *The Heat Is On* and *Boiling Point*: 'The hurricane that struck Louisiana yesterday was nicknamed Katrina by the National Weather Service. Its real name is global warming.'

Katrina was costly, too. The New York-based risk-modelling firm, Risk Management Solutions, estimated the immediate economic damage to be $125 billion. CBS news reported the cost over time could exceed $300 billion. We may never really know the full cost, and much of what was destroyed by the hurricane defies any economic price tag, including the suffering and lives lost. But how sobering are these figures?

Let's re-think the costs of the Three Mile Island nuclear power plant accident, the Alaskan Exxon Valdez oil spill, the Chernobyl accident and a host of other accidents related to energy usage around the world. Even after the news of these leave the media, their consequences, including economic damages running into billions of dollars, continue, while more reports of oil spills and nuclear accidents keep mounting over time.

And let us not close our eyes to the 2010 Gulf Coast oil spill, or get tired of the news. While its catastrophic consequences continue to unfold uncontrollably, with a price tag of billions of dollars, the entire global economy is thrust into an uncertain future. That is a price of oil.

Just the direct cost of the Gulf War in 1991 was estimated to be $61.1 billion.[6] That is in addition to all the lives lost on all sides and the environmental disaster – the long-term worldwide effects of which are yet to be fully determined. Even prior to the Gulf War, it was estimated that in 1989 the US Department of Defense spent as much as $54 billion to safeguard oil supplies from the Persian Gulf.[7]

And now the Iraq War. Based on estimates from the Congressional Budget Office (CBO), by February 2010 the war had already cost US taxpayers $700 billion.[8] That's just the dollar amount for the US alone, while the total continues to go up at the rate of $1000 per second. In a study released in October 2007, the CBO further estimated that the cost of the Iraq War will rise to $1.9 trillion (out of $2.4 trillion for the Iraq and Afghanistan wars combined) by 2017 when counting the huge interest costs due to the combat being financed by borrowed money.[9] The Nobel Prize-winning economist Joseph Stiglitz has assessed it as 'The Three Trillion Dollar War'.[10]

All this resonates so well with what none other than a general, President Dwight D. Eisenhower, said in 1953: 'Every gun that is made, every warship launched, every rocket fired, signifies in the final sense a theft from those who hunger and are not fed, those who are cold and are not clothed.' Simply put, it's the 'Guns or Butter Theory' taught in most introductory economics courses. And, 'To put it on our credit cards with no accountability, with no plan to pay for it, I think is the height of irresponsibility. It will be just one more toxic legacy of this disastrous war we will have to leave our kids to clean up,' said James McGovern, a Massachusetts Democrat who serves on the budget panel. In the root analysis, it is all a part of the cost of oil – the predictably escalating price of depleting nonrenewables.

Only recently has there been a growing discussion on the concept of 'externalities' as they relate to energy costs.[11] What it suggests is that the costs which are related to energy use, yet kept 'external' to the directly charged costs, such as the environmental costs, the subsidy costs, the medical costs, the opportunity costs, the decommissioning costs, the accident costs, the legal costs, the security costs, the war costs, as well as all other indirect or hidden costs, must be accounted for to fully assess the real cost of energy use. Externalities include 'social costs', which fall on society, even though the individual consumer or producer is not directly charged for them, yet may have to pay for them through higher taxes, service cuts, price hikes of oil-based products, inflation, etc., ultimately depriving people of their basic necessities for survival. These costs are significant, estimated to be almost half of the US GDP.[12] Of course, it may be complex, even impossible in some cases, to fully assess such costs.

But compartmentalizing them, diffusing them, subterfuging them, or sheer denial or inability to assess them do not change the fact or solve the problem. The real cost is the *total* cost – the only acceptable economics from an ecological, holistic, global perspective. The total cost is the economic yardstick for sustainability.

Sustainable Economic Development

Rooted in the ecological perspective is the concept of 'Sustainable Economic Development' or 'Sustainable Development'. Much of the impetus for this concept came from the report of the World Commission on Environment and Development, *Our Common Future* (1987), sponsored by the United Nations. It says:

> *where economic growth has led to improvements in living standards, it has sometimes been achieved in ways that are globally damaging in the longer term. Much of the improvement in the past has been based on the use of increasing amounts of raw materials, energy, chemicals, and synthetics and on the creation of pollution that is not adequately accounted for in figuring the costs of the production process. These trends have had unforeseen effects on the environment.*[13]

Economist Herman E. Daly and theologian John B. Cobb, Jr., in their *For the Common Good*, write:

> *We human beings are led to a <u>dead</u> end – all too literally. We are living by an ideology of death and accordingly we are destroying our own humanity and killing the planet. Even the one great success of the program that has governed us, the attainment of material affluence, is now giving way to poverty. The United States is just now gaining a foretaste of the suffering that global economic policies, so enthusiastically embraced, have inflicted on hundreds of millions of others. If we continue on our present paths, future generations, if there are to be any, are condemned to misery.*[14]

The concept of 'Sustainable Development' suggests that, to be sustainable, economic development must consider other factors impacting the future economy, such as the environment. It acknowledges that environmental destruction has a negative impact on the economy, that rampant environmental destruction is a major economic liability and cause of poverty, diseases and violence around the world. The concept owes much to the seminal contribution of David Pearce, economist and professor at University College London, and a modern-day pioneer of environmental economics. Until his unfortunate and sudden death at the age of 63 in 2005, Pearce passionately advanced the notion that the environment has value, and economic growth achieved by devaluing the environment is done at the expense of sustainable development. In his classic *Blueprint for a Green Economy* (1989, co-authored with Anil Markandya and Edward Barbier), originally prepared as a report for the UK Department of the Environment, he wrote:

One of the central themes of environmental economics, and central to sustainable development thinking also, is the need to place proper values on services provided by natural environments. The central problem is that many of these services are provided 'free'. They have a zero price simply because no market place exists in which their true values can be revealed through the acts of buying and selling. Examples might be a fine view, the water purification and storm protection functions of coastal wetlands, or the biological diversity within a tropical forest. The elementary theory of supply and demand tells us that if something is provided at a zero price, more of it will be demanded than if there was a positive price. Very simply, the cheaper it is, the more will be demanded. The danger is that this greater level of demand will be unrelated to the capacity of the relevant natural environments to meet the demand. For example, by treating the ozone layer as a resource with a zero price there never was any incentive to protect it. Its value to human populations and to the global environment in general did not show up anywhere in a balance sheet of profit or loss, or of costs and benefits.

The important principle is that resources and environments serve economic functions and have positive economic value. To treat them as if they had zero value is seriously to risk overusing the resource. An 'economic function' in this context is any service that contributes to human well-being, to the 'standard of living', or 'development'. This simple logic underlines the importance of valuing the environment correctly and integrating those correct values into economic policy.[15]

Pearce sounded radical against the backdrop of the established notion of 'development', which not only ignored the connection between economic development and environment, it even pitted the two against each other – economy or environment, economic development or environmental conservation. But he persisted and succeeded in establishing – through meticulous analysis of environmental values and costs of externalities – both at the academic and policy levels, that 'sustainable development and sustainable growth are interlinked'.[16]

This concept of sustainable development, also emphasized during the UN Conference on Environment and Development (UNCED) in 1992, the Earth Summit in Rio in 1992, the World Summit on Climate Change held in Kyoto, Japan, in 1997, and the World Summit on Sustainable Development in Johannesburg, South Africa, in 2002, continued gaining a greater recognition around the world. During the Rio conference, the Commission on Sustainable Development (CSD) was created under the umbrella of the UN's Economic and Social Council. In the US, the President's Council on Sustainable Development was appointed by President Clinton in June 1993. The council included about 25 representatives from nonprofit environmental and conservation organizations, business and industry, government agencies, Native American nations, civil rights and labour organizations, and charitable foundations. The council defined 'Sustainable Economic Development' as 'development that meets the needs of the present without compromising the ability of future generations to meet their own needs'.

From the perspective of sustainable development, the conventional economics of nonrenewables is the economics of collapse. The commonly held and propagandized notion that nonrenewables are cheaper than renewables is grossly inaccurate, misleading and deceptive. The notion is based on inadequate and incomplete criteria for assessing the real cost. In fact, the real, total cost of nonrenewable energy use may even be out of control and have thrust the world economy into bankruptcy. Even worse, the bankruptcy is denied through passing the cost on to future generations. We are probably the first generation with enough information and knowledge at our disposal to be facing the predicament that, in our convenient, addictive, consumptive, wasteful, compartmentalized and myopic pursuit of a 'good life', we are living off the future of our children. The fact that, for the most part, we close our eyes to that predicament and go on with our pursuit reflects our moral bankruptcy, as well.

Fuelling a sustainable economy

Energy is the lifeline of an economy. The choice of energy will determine whether or not an economy will sustain. The choice, therefore, must be based on some conditions which will continue to energize the economy along sustainable paths. The following are some such conditions, interconnected, which build a necessary infrastructure for the economy to thrive. Therein lie the real cost advantages of renewable energy technologies.

A free and abundant fuel source

The Sun, our basic fuel source, delivers energy through its subsystems of light, heat, wind, water movement and photosynthesis – freely and abundantly. These subsystems are our renewable energy sources. In any part of the world, in a combination appropriate to its location, these fuel sources sustain life. The fuel sources can be tapped through both natural and technological means. This is the basic and sustaining advantage, in contrast to the inevitably rising costs of the rapidly depleting nonrenewables. No matter how much energy prices are artificially manipulated now or in coming years to sell 'cheap' nonrenewables, or how the energy supply is temporarily stretched, the depletion factor – that oil, natural gas and uranium are estimated to be depleted within 50 years and coal around 250 years – is enough to push the prices up, eventually making the nonrenewables unavailable at any price. In contrast, the Sun, at least for its lifetime of another 5 to 10 billion years will continue to supply the renewable fuels – freely and abundantly!

Environmental advantages

It is a blessing that the main push behind renewable energy development has been, and continues to be, the environmental movement. This is the best safeguard we have against costly economic consequences of environmental destruction. The link between technology, environment and economy must be continually emphasized and scrutinized – in terms of both short-term and long-term implications. While each technology must be examined individually, in all its aspects and phases – such as extraction of resources, manufacturing, operation, disposal, recyclability – studies suggest that renewable energy technologies, in general, far exceed nonrenewable energy technologies in terms of environmental advantages. It is well acknowledged by the industry that the appeal of these technologies is profoundly contingent on their technological and economic advantages as well as environmental advantages. The required level of environmental control is borne by the manufacturer and becomes a factor in the selling price, which, consequently, becomes a built-in mechanism to check against hidden and external costs.[17]

The downward trend of technology costs

As with computers and televisions, breakthrough developments in efficiency, scale of production and manufacturing processes have led to declining costs of renewable energy technologies over the years. The direct costs of photovoltaics and wind energy have come down by 60 to 80 per cent since the early 1980s. Very few products or technologies can claim that steep a downward cost-curve. However, a sudden increase in demand without adequate production and supply always leads to price increases – as we have witnessed lately. But it is more like a bump on the road which will not last. Many more manufacturers around the world are entering the industry. Current manufacturers are planning on boosting production. Development of diversified raw materials, mass production and more efficient as well as newer technologies, in massive volumes, will continue to emerge and the downward trend of costs will continue.

Remote and urban applicability

Renewable energy technologies are economically advantageous for both remote *and* urban settings. The revolutionary advantages of remote or decentralized applications are better recognized. Recent advancements, along with a downward cost, have made renewable energy technologies advantageous for urban applications, as well. That, too, comes with a revolutionary potential.

Remote advantage

For remote or decentralized applications beyond the reach of conventional power lines, a PV, wind turbine or hydroelectric system – appropriately matching the energy source – is well-proven to be the least-cost option for electricity generation. The modular systems can be installed as stand-alone or hybrid systems, in the latter combining two or more different types of systems.

The scope for remote applications abounds – from space satellites to navigational buoys for ships off the New England coast, from a rural home in Bangladesh to a water pumping station in the Dominican Republic, from a telephone booth in a Saudi Arabian desert to a village clinic in Zaire, from a street-light in Sri Lanka to a forestry guard post in Armenia, from a roadside tea stall in India to a research station in the Arctic, from a village community centre in Nepal to a telecommunication station in Arizona. A wide variety of remote sites are dispersed throughout the world.

The most dramatic scope for remote applications of renewable energy technologies lies in the developing countries. According to a UNEP report, at least 1.6 billion people in the developing world do not have access to electricity.[18] These are people who live in the remote areas of developing countries, where up to 70 per cent of the area is beyond the reach of the existing power lines. And most of the countries in the world fall into the 'developing' countries category. Even in places where people have the access technically – meaning they have power lines – the actual supply of electricity there is highly insufficient and unreliable. This means, for most of the world renewable energy technologies are already the least-cost options for electricity generation and the most cost-effective, sustainable and achievable solutions toward a renewable energy transition.

Solar electrification also carries with it many other immediate economic advantages for developing countries, such as expansion of cottage industries, night-time vocational training, use of power tools, opportunities for small traders, shop-owners and other entrepreneurs, and higher productivity through better and healthier lighting – compared to lighting from commonly used kerosene lamps. Small stand-alone PV systems have also proven to be more cost-effective against kerosene lamps over time.[19] The socio-economic benefit of poverty alleviation through solar electrification is universally recognized. And so are the benefits of healthcare, education and recreational opportunities, whether at the household or community level, for both adults and children.

Urban advantage

Cities and towns are made up of buildings, and buildings have roofs and walls. If sunlight reaches them, they are prime spaces for generating heat and light.

In the heart of Boston is the State Transportation Building, a gigantic 82,000m² combination of office and commercial space. Installed on its roof is a solar thermal system which, for 20 years, has been supplying over 80 per cent of the building's hot water needs. In 2008 the Boston Red Sox's Fenway Park stadium installed a solar hot water system which supplies 37 per cent of its hot water needs. Multi-storey residential buildings, offices, hotels and other institutions are beginning to realize this tremendous urban advantage and examples like these are popping up in cities across the country and around the world.

One of the breakthrough developments in PV technology is the 'grid-connected' system. It is revolutionizing the way electricity is generated in and for urban settings with conventional power lines. Electricity is generated by renewable energy systems, such as PV or wind turbines, injected into the grid, and then distributed through the power lines, supplementing the power generated through conventional sources. Just think how much electricity could be generated by putting up some PV modules on the roofs or walls of city buildings around the world!

No place in the world can be considered more urban than New York's Times Square. At the heart of it, a 48-storey skyscraper has a 'PV skin' which replaces the traditional glass wall, generating electricity to meet partially the needs of the building. The PV-integrated architectural glass is translucent, so – like a window – one can see through it. The building also incorporates features such as fuel cells and daylighting, showcasing urban applications of renewable energy technologies.[20]

Kyocera Corporation's headquarters, a 20-storey skyscraper in Kyoto, Japan, has PV modules installed vertically on one of its side walls, along with other 'green' features.[21] There are several PV-integrated skyscrapers in Bern, Switzerland, in which PV modules replaced other materials for covering the façades and therefore served as the building materials themselves.[22] These examples are just the beginning and there are many more roofs and walls around the world – sitting idle. So, instead of just sitting idle all day, they could be doubly productive as a roof or a wall and as a power plant! As the California utility PG&E (Pacific Gas & Electric) put it in a newspaper promotion for renewable energy, 'Green is giving your roof a day job.'

This job with many openings – with the appearance of vacant roofs and walls – needs to be urgently filled! Especially during an economic recession, compounded by an ever-costlier and worsening energy crisis, that's good news. Around the world in areas with a grid, grid-connected solar systems will be one of the fastest, simplest, most cost-effective and environmentally advantageous ways to add both generating and transmitting capacities to the grid. With multitudes of untapped urban areas around the word, the potential of such applications to facilitate the transition to the renewable energy path is truly revolutionary.

Turnkey transferability

Especially when the energy sources, human resources and most of the manufacturing materials – three key ingredients in technological production – are locally available, to move on to actually manufacturing the technologies locally, instead of importing them, may make the most economic as well as political sense. It fuels the economy, creates 'green-collar' jobs and helps achieve energy independence. Sometimes a country may import just the cells to take advantage of the economy of scale in cell manufacture elsewhere, or wait until its infrastructure is ready to take on that aspect of the manufacturing process while it manufactures the rest of the components locally. It has been well demonstrated that by combining turnkey transfer with local innovation, modules and other components can be manufactured in countries around the world that are not only among the best in quality, but the most cost-effective, as well. That is a 'stimulus package' which keeps stimulating the economy!

There is a growing global trend of turnkey transfers of renewable energy technologies. Several PV manufacturers supply turnkey production lines. They can be contracted for either a part or the entire process of planning, training local expertise and manufacturing PV modules locally from scratch. Then they turn over the key to the contractor and leave – hence, the phrase, 'turnkey transfer'. For example, Spire Corporation, 'The Turnkey Factory Company', a Massachusetts-based PV manufacturing company and a pioneer in the industry, has transferred turnkey facilities – the complete line or some segments – to 50 countries worldwide.[23]

The manufacturing process requires energy, which can be a challenging, if not prohibitive, factor for many developing countries. The PV-manufacturing plant of the Solarex Corporation in Frederick, Maryland, set a classic example of a solution to the challenge. The plant was built in the 1980s to receive all the energy it needs from its 200kW of PV modules on its south-facing sloping roof to manufacture 5MW of new solar cells each year. For this reason it has been called a 'solar breeder'. It became a prototype for countries that cannot afford, or simply do not have, the electric power needed to manufacture PV modules.[24] In 1999 Solarex merged with BP Solar, first becoming BP Solarex, then BP Solar. BP Solar in Frederick has now grown to be one of the nation's largest fully integrated PV-manufacturing plants. But this need to rely on renewable energy, not only in applications of technologies, but also for the manufacturing of technologies itself, is a transitional necessity, no matter in which country the manufacturing is taking place. An inspiring example is set by Germany's Freiburg-based company, Solar-Fabrik: 'The company operates an emission-free plant that gets its power from its own photovoltaic modules and a biomass cogeneration unit. The only emissions from this plant are related to purchased materials.'[25]

As it is with PV, turnkey transferability is a common advantage for a number of other renewable energy technologies, including a variety of wind turbines, hydroelectric and solar thermal systems. The trend is growing there, too.

Alleviation of debts and deficits

Mounting foreign debts have become cyclical as well as worsening problems for many industrially developing countries where the fuel import bills amount to as much as 70 to 90 per cent of the total export earnings. Funds needed for many essential purposes go to pay off foreign debts, often only to qualify for additional loans – in reality, more debts.[26]

'Solutions' are at best treating symptoms if economies continue to rely on nonrenewables. As the costs of nonrenewables rise while the energy needs of the economies grow, the fuel bills grow too. The crisis is further complicated and worsened by the associated externals and social costs impacting the economy, such as climate change and energy insecurity.

This affects the lender adversely as well. Let me illustrate with a reference to banking. It may be somewhat simplistic, but illustrative nevertheless. The bank, the lender, lends money to benefit itself from the interest the borrower pays to it. As long as the borrower can pay the interest, while only fractionally amortizing the capital, the bank is fine and profiting – and that's just what the bank prefers to get from the borrower, rather than the return of the capital. That is why the interest amount is much higher in the beginning than the amount by which the capital amount is being amortized. Some banks even charge a penalty fee for paying off the loan too early. Now, if for some reason the interest payment itself becomes prohibitive for the borrower, the bank may lend additional money to bail out the borrower so that the borrower may resume paying the interest which the bank survives or thrives on.

However, if the process continues, that is the borrower is increasingly defaulting on the interest payments because that amount itself has reached a proportion beyond their ability to pay, then lending more money is no longer a solution for the bank or the borrower. The bank itself is in jeopardy because, along with losing its income from the interest payments, its assets – the capital lent out to earn that interest – are at risk of being lost. The greater the proportion of its total assets that the bank has lent out in its desperate attempt to salvage the default, the greater is the risk and the hit to which the bank is subjected. Depending on the value of the loan collateral, the declaration of bankruptcy by the borrower may only add to the lender's liabilities. Of course, banks operate with all kinds of safeguards around their assets and investments. Still they collapse – as happened during the banking crises of the 1980s and the continuing housing and financial crisis since 2007. Moreover, the costs of those safeguards fall on the overall economy of which the bank itself is a part.

So, there is a limit as to how far such safeguards can be extended, without adversely affecting both the bank and the economy as a whole. Much of the debtor-creditor relationship between the industrially developed and developing countries can be explained in this way. The relationship between the World Bank and the economies of the developing countries are similar. It is a trend, unless some fundamental changes can reverse it, of eventual global economic collapse – the signs of which are becoming all too evident these days, and the effects of which are causing much too much pain to suppress.

Especially since energy costs make up such a high percentage of developing countries' foreign debt, turning to renewables is one essential step towards breaking out of the vicious cycle of dependency. The more a country can rely on renewable sources, the more its economy is freed of fuel costs, thereby alleviating the major source of foreign debt. A country can develop or revitalize technologies appropriate to its renewable energy sources. Moreover, it can take advantage of the turnkey transferability of the renewable energy technologies to empower itself. Or it may at least begin to manufacture some of the components locally, thereby investing that much of its energy costs in its own economy, invigorating it as a result. Even if the initial capital needed for such investments comes from external sources, the situation is fundamentally different from reinforcing the debt cycle. Over time it can result in even more than breaking the debt cycle and paying off foreign debt. Investments in renewables are also investments in uplifting the economy towards greater self-reliance and sustainability, both for national and global economies.

Turning now to the industrially developed countries, the economic prosperity of these countries has been primarily fuelled by nonrenewable energy sources. Energy being the lifeline of an economy, the rapid depletion of these sources alone is leading these economies to their eventual collapse – with catastrophic impacts on the world economy (including the developing economies so dependent on them). In the meantime, energy prices are heavily subsidized and kept artificially low at the expense of a growing number of essential social, human and environmental needs and services in the present, with even more critical implications for the future. Neglected externalities or social costs – 'total cost' accounting – are rapidly adding up to put future generations in debt even before they have been born. In reality, this constitutes the transfer or displacement of the bankruptcy of the present economies into the future.

For some of the industrially developed countries imported oil bills account for a major and ever growing portion of their national expenditure. The US imports 70 per cent of the oil it consumes, paying over $700 billion annually, and its net oil imports represent more than 53 per cent of the total US merchandise trade deficit. It is complex to single out an item as a cause of trade deficit, but oil is significant enough a variable for analysts to identify that increased oil imports and the rising price of oil cause the trade deficit to surge.[27] In turn, such surges also add to the ballooning national budget deficits and debts. Unchecked, they indicate symptoms of a collapsing economy.[28]

Fifty-four per cent of the US federal budget for Fiscal Year 2009 was spent on the military (18 per cent on past military and 36 per cent on current military expenses).[29] So much of the spending – under various allocation labels – is related to dependency on, and securing the supply of, oil. How long do we go on crippling our economy and escalating debts and deficits, before transitioning to the renewable path?

Reliance on local renewable resources – a secure and self-reliant way to revitalize national productivity – is necessary to alleviate national expenditures, debts and deficits, thereby reversing the collapsing trend.

Equity and justice

While fuelling global industrialization, the nonrenewable energy path has also been fuelling the global inequity between the industrially developed and developing countries. With 5 per cent of the world's population, the US consumes 24 per cent of the energy produced in the world. The developed countries as a whole, with 20 per cent of the world population, consume 58 per cent of total energy. According to the Worldwatch Institute, the seven largest industrial nations – with a total of barely 10 per cent of the world's population – together account for more than 40 per cent of the world's consumption of fossil fuels, most of its consumption of metals, and a major percentage of its consumption of forest products and industrial materials.[30]

The transition from the nonrenewable to the renewable energy path can equalize – and therefore, in the true sense, democratize – this fundamental and pervasive means of control over the global economy: the energy supply. The critical link between equity, energy and democracy needs to be understood and acted upon. And the Sun will be there to fuel the equalizing, democratizing process. An added advantage: most of the developing countries, located in sunny tropical or semitropical regions, are phenomenally rich in their renewable energy sources – practically 'untapped energy mines'.

A critical awareness of inequities, exploitation and environmental victimization is growing in the developing countries. It's fuelled greatly by the knowledge of the causes and catastrophic consequences of climate change. There's a growing awareness that the developed countries, the main beneficiaries of utilizing fossil fuels, are also most responsible for climate change. At the same time, the developing countries, the far lesser beneficiaries of utilizing fossil fuels, are also among the least responsible and worst victims of climate change. Land disappearance under water due to rising sea levels, environmental refugee crisis, civil unrest, diseases, droughts, salinity and other contamination of water and cropland, and deaths, to name a few, are exploding realities. According to the United Nations, 'Over the next decades, it is predicted that billions of people, particularly those in developing countries, face shortages of water and food and greater risks to health and life as a result of climate change.'[31]

But it's not just about the future. Climate change is happening right now, victimizing millions of people and other species around the world and causing irreversible damages to the environment. For example, more frequent cyclones and sustained droughts are present realities that are consistent with what scientists have been predicting and equating with climate change. When a cyclone hitting Bangladesh or a drought in an African country is reported on television, we see the actual images of a suffering humanity and a devastated Earth. Even though those images last only for a few fleeting moments, the suffering and devastation continue. This means every greenhouse-gas-emitting power plant that goes up anywhere in the world carries with it the decree of death and destruction for many who are disenfranchised, who'll not benefit from it, but become its innocent victims. This far outweighs the supposed benefit it promises for a few. The sufferers can no longer be categorized merely as 'victims of nature' – to be treated with condescension and charity. Instead, there is now an awareness which stems from legal and human rights perspectives which demands – through action, not just promises – compensation; support for mitigation, adaptation and renewable energy technologies; migration rights; and equitable justice. And this viewpoint demands that the developed countries curb their own consumption patterns responsible for climate change. The voice has been raised – and has become louder – in the world summits on environment and sustainable development, and was heard the loudest at the UN's Climate Conference in Copenhagen in December 2009. This is inevitable and morally justifiable.

How far will the pendulum swing? Nobody can tell. There's a long history of inequities and injustices to contend with. It would be naive to say that inequities and injustices in the developing countries are all externally imposed. Most monarchies, feudal states and chiefdoms thrived through inequities and exploitation of the masses, as do most of the present-day 'liberated' nations under their corrupt political leaderships and hierarchical social stratifications. They simply can't be justified. However, what we are focusing on here are the inequities and injustices which have been inherent in the ideological progression of 'colonizing' to 'civilizing' to 'developing', which have been façades for exploiting the vast nonrenewable resources around the world in the interest of the few colonial powers to the 'civilized' countries to the modern-day 'developed' countries.

Climate change is unravelling a piece of that history fuelled by a critical consciousness. When Nature's fury combines with human fury, it's a recipe for a Perfect Storm. It spares none. But I hope that we can still avoid that. And we need to avoid that in common interest and for the common good. However disproportionate the victimization between developing and developed countries may seem now, eventually the catastrophic impact of climate change will know no borders. The unusually changing climate patterns, record-breaking snow storms, rainfall and droughts across North America and Europe are indicating what's yet to come. Mohamed Aslam, the Environment Minister of the Maldives, an island nation in the Indian Ocean whose virtual existence (according to the IPCC) is threatened by the rising sea levels, says it well: 'The best we can do is tell the world that what is happening to us can happen to you tomorrow. The big countries must see their future reflected through us.'[32] With the alarming growth of evidence of the effects of climate change – around the world and across developed and developing countries – the 'tomorrow' is already here, today. Acting on solutions, the Maldives has announced a plan to become the world's first carbon-neutral country by generating all its electricity with photovoltaics and wind turbines.

To press for global action – not just words – the Maldives' initial dramatic decision to not join the Copenhagen summit and hold an underwater conference on climate change at home sent a clarion call for action on behalf of the entire world community. Unfortunately, keeping with the tradition of international conferences which breed more conferences and committees than solutions, the Copenhagen conference turned out to be the most recent enactment of another self-congratulatory, elitist and foot-dragging drama, devoid of substance and action but full of loopholes and empty promises, presented on the world stage by the leaders of the developed nations. However, the voices demanding action, too, have continued to get louder, and so has the determination to take action locally at the grassroots and community levels around the world.[33] An action-oriented global network of grassroots solidarity is emerging. 350.org, led by Bill McKibben, is an excellent example. That, indeed, is a very hopeful trend.

So, together, let's choose the path of equitable solutions over global catastrophe. Just as the nonrenewable energy path has fuelled the scenario, a sincere and mutual commitment to equity and justice along with earnest collaboration to expedite energy self-reliance through the renewable energy path for every country – developed or developing – will play a key and redeeming role in the transition of the world as a whole toward a more sustainable future.

Peaceability

As nonrenewables become increasingly scarce and the economy becomes increasingly dependent on imported oil or other nonrenewable energy sources, global tension and insecurity rise. The oil crisis of the 1970s and the Middle East wars demonstrate this phenomenon.

The situation did not get to be that way overnight or by accident. Oil companies – British Petroleum (BP, formerly the Anglo-Persian Oil Company), Exxon, Shell, and Aramco (an oil consortium of US oil companies) have been exploring and cutting deals in the region since the early 1900s. The stage was set and the drama continued to unfold. As the vast oil reserves – 'The Prize' – were being discovered in the Middle East, a meeting between King Ibn Saud and President Franklin D. Roosevelt aboard an American ship in the Suez Canal in 1945 laid the groundwork for the US – particularly its oil corporations – to come out as the winner. It was a simple agreement: the Saudi monarchy will permit the US to exploit oil and, in turn, the US will protect the Saudi monarchy.[34]

Pursuit for oil in the region continued – only more aggressively and compounded by Kuwait and Iraq factors – escalating to today's situation which will have no winners, only victims. However justified or politicized, and no matter what other causes or solutions are sought while ignoring the root cause, that path of occupation, exploitation and retaliation will lead us to nowhere but a dead end with only more human, environmental and economic costs on all sides. Cogently warning about these inevitable consequences of the oil path, while proposing clean energy alternatives, Paul Epstein and Chidi Achebe conclude: 'Once called "The Prize" by Daniel Yergin in 1991, oil has become "the curse".'[35] Look around the world – from Nigeria to Sudan (Darfur, to be more specific) to Iraq to Colombia to Azerbaijan to Georgia to Papua New Guinea to Bangladesh – and see how relevant this statement is to any place where oil or any other nonrenewable energy resources are found.

Moreover, due to the centralized, concentrated power necessary for controlling nonrenewable resources, nationally and globally, the warring scenario will also be marked by accelerating militarization and totalitarian conditions, demanding from us to surrender – in the name of false security – whatever dwindling civil liberties we have left to cherish. Recent years bear ominous signs. In *Real Security*, Richard J. Barnet wrote 20 years ago but could have written today:

> *The decision to invest a trillion dollars in the military rather than in a crash energy-development programme to reduce a dangerous dependence on foreign oil is a prime example of increasing the nation's vulnerability by piling up hardware and expensive military bureaucracies. The hardware cannot produce energy; it consumes energy. It cannot assure access to energy, because there is no effective military strategy to assure the flow of oil through a system vulnerable to sabotage.*

> *Useless military forces pre-empt investment funds, public and private, that could be used to develop alternative national-security strategies appropriate to the new century which we are about to enter. Increasingly, national power comes out of innovative minds rather than the barrel of guns.*[36]

The fuel war has also refuelled the proliferation of nuclear power and nuclear weapons production. The two are intrinsically linked. Two of the byproducts of nuclear power production are highly enriched uranium (HEU) and plutonium, the main ingredients for nuclear weapons including bombs. A typical nuclear reactor produces about 250kg (550lb) of plutonium per year – enough for 25 to 50 weapons. As of early 2007, the global stockpile of HEU reached up to 200 tons and the global stockpile of separated plutonium reached about 500 tons.[37] Behind the façade of secure storage, these ingredients are hotly pursued items through secret deals and theft. The vast and proliferating arsenals of nuclear weapons around the world are a testimony to the chilling truth that the so-called 'peaceful atom' in reality has been far from peaceful. As Robert Jungk puts it: 'The splitting of the atom moved man into a new dimension of violence. What began as a weapon against one's enemies now threatens the hand that wields it. In substance there is no difference between "Atoms for War" and "Atoms for Peace".'[38]

Einstein reminded us that we cannot prepare for war and peace at the same time. Around 30 years ago Amory and Hunter Lovins, co-founders of the Rocky Mountain Institute, which is a living example of sustainable design, and pioneering advocates of sustainable development, warned the world about the inherent and terrifying link between nuclear power and nuclear weapons, between energy and war, and the critical choice we face between energy for war and energy for peace.[39] We must address these links and choices now, especially that nuclear power is gaining momentum, propagandized as a solution to climate change. The possibility of a sustainable future depends on it.

The sustainable economy: building a new foundation

What we need is a long-term vision of sustainability, with urgent actions rooted in that vision. As in Gandhi's holistic notion of an 'inviolable connection between the means and the end, as there is between the seed and the tree', the urgent actions and long-term goals have to be integral elements of a continuous whole.

Renewable energy – with its vast sources of light, heat, wind, water movement and photosynthesis – has the potential to revolutionize the global economy. Fuelled by these renewable sources, the choices of many ingenious and practical solar technologies, including photovoltaics, wind turbines, hydroelectric and solar thermal systems, biogas and solar cookers are within our reach right now. Appropriately chosen and combined, these offer sustainable solutions for rural as well as urban settings around the world. A wide range of applications of these technologies have proven their economic, environmental and political advantages. Many more promising options, such as fuel cells, biofuels and geothermal systems, already with some proven results, are on the horizon.

The transition will not be easy, especially because of our deep entrenchment in the nonrenewable path, fuelled in particular by the energy-guzzling consumption practices we have uncritically come to equate with the 'good life', and the powerful vested interests that guard and reinforce the entrenchment. Addiction to oil, reinforced by both ignorance and denial, runs pervasively through the industrial culture, while the demand for energy escalates both in the industrial and industrially developing countries around the world.

India and China, the world's fastest industrially developing countries which are aggressively pursuing the nonrenewable path (far outstripping their own otherwise impressive investments in the renewable path), pose two of the greatest emerging environmental threats to the world. In the long run, projected according to the sustainability yardstick, they are destined for their own economic – and environmental – collapse, as well. China and India are also two of the countries in the world with the most enduring and profound wisdom of sustainability interwoven in their cultural-spiritual fabrics. Let us hope, fuelled by the growing environmental movements in these countries, this wisdom prevails. In Indian mythology, the snow-capped, unconquerable and mysterious Himalayas are the abode of the gods who have been sustaining creation from time immemorial. In turn, humans have looked upon the Himalayas with the utmost reverence. There couldn't be a more serious warning against climate change than the accelerated melting of the Himalayan glaciers due to the Earth's rising temperatures.[40]

But do not let India and China divert our focus from the industrially developed countries. While India and China are emerging as the greatest environmental threats, the industrially developed countries are already that, having caused damages worldwide that we are just beginning to experience. All the scapegoating and finger-pointing at India and China will amount to little more than hypocrisy unless the developed countries themselves take measures to curb their own rampant, environmentally destructive practices and quest for profit maximization by taking advantage of China and India's cheap goods and services and thereby fuelling environmentally irresponsible production. To heal the addiction, we still have to find the right doctor or, at least, a global policy that can curb globalizing consumerism and greed, a policy that will treat both the addict and the pusher – jointly and equitably.

The US Department of Energy, in its *International Energy Outlook* (1998), predicted that global energy demand will climb by 86 per cent between 1990 and 2020. The Energy Information Administration (EIA) of the US government projects world energy consumption to nearly double between 2004 and 2030. The hi-tech revolution around the world is consuming power like there's no tomorrow. In the US, computers and other electronic gadgets account for nearly 10 per cent of total power consumption.[41] Electricity has been called the 'oxygen of Silicon Valley'. Silicon Valley is now facing a growing shortage of oxygen. Merely outsourcing Silicon Valley operations to India, China and wherever else, relying on the same source of oxygen, does not solve the problem, it only globalizes it.

In the face of this growing energy shortage from nonrenewables, barely 7 per cent of the world's energy is generated from renewable energy technologies. Projections for such generation by 2030 do not exceed 10 per cent. The worldwide renewable energy technology business, excluding large hydros, is expected to reach $35 billion in 2013, up from $7 billion in 2004.[42] However impressive this is in isolation, it does not come close to reversing the trend of further entrenchment into the nonrenewable path.

Some analysts have suggested that, as nonrenewables become scarcer and costlier and renewable technologies become more cost-effective, market dynamics will turn in favour of renewables.[43] This is grounds for optimism. However, there are real threats and obstacles to such possibilities, which we need to remove urgently through public policies so the market dynamics can play out that role. It is true that profitability attracts private investors, with or without governmental support – that is the law of the marketplace. It is not a surprise, therefore, that with the growing demand for renewable energy products, the number of investors is growing, too. Venture capitalists are migrating into the renewable energy technology territory like a new 'Gold Rush', or even better, a 'Green Rush'. Investments in 'cleantech' are at an all-time high. Nor is it a surprise that most of the major investors in renewable energy technologies are nonrenewable energy corporations or their subsidiaries. Oil and auto giants – BP, Shell, Atlantic Richfield (ARCO), Chevron, Texaco, ExxonMobil, Daimler, Mitsubishi, GM, Ford – they're all in it. BP in billboards and magazines around the world declares itself as 'Beyond Petroleum'. 'Ignoring alternatives is no alternative,' Shell admits, then goes on to advertise its 'commitment to sustainable development, balancing economic progress with environmental care and social responsibility. So with real goals and investment, energy from the sun can be more than just a daydream.'

But there's another side to how these corporations actually behave. Investment of these corporations in renewables is a pittance compared to their investment in nonrenewables. Their influence through lobbyists at the governmental level reflects a wide discrepancy in the way the government subsidies are distributed between nonrenewable and renewable research and development, favouring nonrenewables.[44] Their oligopolistic control of the economy, politics and policymaking overpowers the free-market dynamics of supply and demand. Their ruthless pursuit of the increasingly scarce, therefore also more profitable, nonrenewables around the world raises questions about their social or environmental responsibilities in any fundamental sense. The record high oil prices – crippling the economy down to household budgets – lead to a record high profit for the oil corporations. Analyse the equation. And that motive – maximizing profit from nonrenewables – is what rules the corporate policy and practice.

With an investment of $124 billion between 2007 and 2012 to produce oil from tar sands in Alberta, Canada is poised to become the 'new Saudi Arabia'. As global warming melts the Arctic, countries including Russia, Canada, the US and several others are rushing to hoist their flags and start digging, claiming a piece of the 14.2 million square kilometres of the region, estimated to hold 25 per cent of the Earth's oil and natural gas reserves. Propagandizing the excuse of combating climate change, while suppressing any mention of the host of risks and consequences associated with nuclear power, the nuclear industry is pushing for a comeback. It has found an ally in the International Energy Agency, a Paris-based intergovernmental organization founded by the Organisation for Economic Co-operation and Development (OECD), which proposes building 1400 nuclear power plants worldwide in the coming decades. GEI Consultants, while proposing a national energy plan to include exploitation of all potential power sources, and dubbing nuclear power plants a 'green technology', an oxymoron, advocates building 40 new nuclear power plants in the US by 2035.[45]

For all these reasons, the corporate responsibilities seem more of a deceptive 'greenwash' and far less a priority compared to their short-term goal of maximizing profit from nonrenewables. Left to market dynamics, that scenario is much more imminent than an actual transition to the renewable path.

To bring about the transition, a new political will – representing the true public will, not the corporate will – must be asserted and new policies need to be implemented. The policies must require converting their investments and subsidies in favour of renewables. Using the nonrenewables only as transitional fuels and a moratorium on investments in further entrenchment into nonrenewables, combined with a crash infusion of investments, incentives and programme implementations in renewables, will be necessary. The corporate domination and practices of profit maximization for a few at any cost must be checked by public oversight, out of which a socially and environmentally responsible public-private partnership (PPP) can flourish. These measures have to be executed with the utmost urgency. As the concern over the dire consequences of climate change grows, some environmentalists and politicians have said that combating climate change successfully will require putting out policies and programmes with a priority and magnitude comparable to the Manhattan Project, the US programme (1942–46) to build the atomic bomb. We will need a Manhattan Project-type priority to make a transition to the renewable energy path before it is too late. Also, we need to come to the quick realization that, in terms of priority and magnitude, the aggressive pursuit of nonrenewables – overtly and covertly – by corporations and state monopolies, such as Russia and China, are already beginning to resemble the Manhattan Project. Only a socially and environmentally responsible public-private partnership will be able to redirect that pursuit toward a renewable energy path.

Not exhausting the nonrenewables does not mean those resources are wasted. The Earth, as a living organism, has its own reasons and need for those resources. The modern scientific hypothesis about the Earth as a living and evolving organism, Gaia, which sustains life through a process of interdependence of elements within itself as well as with elements of the larger ecosphere, is also ancient wisdom.[46] Conserving and renewing those resources is a biological necessity for the Earth itself. For the Earth to live, to thrive, to sustain itself, it must be able to rely on its life-sustaining elements. As we consume the Earth's resources in order to sustain, 'Hymn to the Earth' in the Vedas, the ancient Indian spiritual scripture imbued with wisdom and recorded around 1500 B.C., reminds us of this responsibility:

> *Whatever I dig up of you, O Earth,*
> *may you of that have quick replenishment!*

Our own human ancestors had the holistic wisdom to understand the meaning of sustainability that we need to live within the limits of the Earth's natural resources and our renewable capacities. Rooted in that wisdom was a foundation of what we have come to conceptualize as 'Sustainable Development', in which economic utilization of natural resources and spirituality were fused. Yet, 3500 years later, we find ourselves the farthest away from that wisdom, posing the greatest threat to the health and survival of the Earth itself. World Wildlife Fund's *Living Planet Report*, released in October 2006 says:

> *The* Living Planet Report 2006 *confirms that we are using the planet's resources faster than they can be renewed – the latest data available (for 2003) indicate that humanity's Ecological Footprint, our impact upon the planet, has more than tripled since 1961. Our footprint now exceeds the world's ability to regenerate by about 25 per cent.*

> *The consequences of our accelerating pressure on Earth's natural system are both predictable and dire. The other index in this report, the Living Planet Index, shows a rapid and continuous loss of biodiversity – populations of vertebrate species have declined by about one third since 1970. This confirms previous trends.*

> *The message of these two indices is clear and urgent: we have been exceeding the Earth's ability to support our life-styles for the past twenty years, and we need to stop. We must balance our consumption with the natural world's capacity to regenerate and absorb our wastes. If we do not, we risk irreversible damage.*

> *We know where to start. The biggest contributor to our footprint is the way in which we generate and use energy. The* Living Planet Report *indicates that our reliance on fossil fuels to meet our energy needs continues to grow and that climate-changing emissions now make up 48 per cent – almost half – of our global footprint.*[47]

So, it's time to take this warning – and responsibility – seriously, both for the sake of our own survival as well as the Earth's well-being, and the two are intrinsically linked.

But let us assume, on the other hand, that the Earth survives the depleting assaults – for the Earth has also proven itself to be immensely resilient, intelligent and powerful. Would it then go on accommodating humans as one of its inhabitants? Or – as an organism tends to do to an element which behaves 'badly' or threatens its survival – would it declare 'Enough!' and let the human species drive itself to its own extinction? Or, at least, give it a thorough cleansing? If we truly valued the Earth's organic integrity and its 'Earthright' – if you will – to sustain itself, then the human assaults on it would be justly considered punishable crimes. We must challenge the notion of ownership and rights of any nation or corporation to exploit and exhaust for short-term gain for the few the Earth's resources that are sustaining elements of the Earth itself, the home for *all* its inhabitants in the present and into the future. If we remain blind to that clause of environmental justice and continue with our assaults instead, will the Earth be compelled to be the ultimate judge in its own defence?

Either way, these are critical questions and an urgent call for building a new foundation for sustainability.

The transition is a process, but a process that will require us to act with the utmost urgency, in every possible way. It will not be easy. But the challenges, complexities, costs, even sacrifices that are inevitable to the transitional process are minimal compared to the economic costs and other consequences of not making the transition. The British government report's projection that global warming could cost the world economies up to 20 per cent of their gross national product, alarming as it is, still reflects merely a fraction of the total cost. Despite all the mis-education, brainwashing, confusion, denial, Big Oil manipulation and weapon industry propaganda, the truth about the critical consequences of climate change in Al Gore's *An Inconvenient Truth* is increasingly inescapable, and far worse than anything measurable by a loss of GNPs.[48]

Maximizing the use of renewables while conserving and utilizing the nonrenewables with the utmost efficiency and as *transitional* fuels (with a moratorium on further entrenchment into the nonrenewable path) is the key to the transition. Every individual, community, national and international action counts. The diverse nature of the renewable energy technologies offers enormous possibilities for such actions, anywhere in the world.

Just as importantly, we will need all the people with their expertise in the conventional energy technologies to participate in, and even guide, the transitional process. The conventional power plants will still need to be operated and maintained efficiently while they are still in operation and energy services have to be delivered reliably, combined with proactive and collaborative steps towards the transition. The transition will mean not mere unemployment of those expertises; rather, transformation and renewal of these expertises into essential contributors in the vital transformation of the economy towards sustainability.

Contrary to a common misconception, the renewable energy path is not about giving up a lifestyle. Instead, it is about giving up an unsustainable 'deathstyle' while renewing a lifestyle which can be sustained through generations. It is about choosing to live a life beyond instant gratification, over consumption, waste, short-sighted pursuits and a compartmentalized existence and regaining our essential interconnectedness, responsibility, wholeness and sanity. It is about learning to live more simply and richly with available resources with fulfilment, equitability, universality and the abundance of the renewability of resources. It is about giving up reinforcing the growing forces of destruction which, sometimes violently, and most of the times cancerously, take away from us the very best achievements of our past and present, and the achievements lying ahead in our future. What we have to give up is what is not going to be possible to hold on to much longer anyway, and will lead us to nowhere but dead ends. And it is about saving ourselves and other species from the growing threat of a depleted Earth, poisoned air-water-soil, global warming, acid rain and nuclear contamination. And it is about not having to go to war over oil.

What we need is a breath of life into our failing life-support systems that are rapidly running out of fuel and poisoning and destroying our ecosphere as well. The Sun has breathed life into these systems for billions of years. If we allow ourselves to receive it, it will do so for billions of years to come. The Sun continues to shine upon us: we can either keep our eyes closed – even while facing it – and pretend that it's dark, or we can open our eyes and see the light.

Ultimately, it's a moral challenge that we face. It is about the values which we want to live by, and the choices we make. And no choice is more crucial than the choice of the energy path in determining the future we are going to face. It is a choice we face – now!

5
The Renewable Revolution:
Turning Vision into Action

I'm ten years old and very worried about our environment. I wish I could feel free to breathe the air I do breathe, swim in the water I do swim in, look at the ugly, diseased, or burnt trees that were once beautiful. I sometimes wonder if you really do anything about it? Why, and you ask what do you mean why? Well, I mean, why just stand (or sit) there reading my letter!! DO SOMETHING!!!   A concerned fourth grader, Kristie Sue Houch[1]

We have less than 10 years to halt the global rise in greenhouse gas emissions if we are to avoid catastrophic consequences for people and the planet. It is, simply, the greatest collective challenge we face as a human family.
United Nations Secretary General Ban Ki-moon[2]

Now that you're here,
The word of the Lorax seems perfectly clear.
UNLESS someone like you
cares a whole awful lot,
nothing is going to get better.
It's not.

Dr Seuss, *The Lorax*[3]

Half a century ago, in *The Coming Age of Solar Energy*, D. S. Halacy wrote: 'As we stand on the threshold of the age of solar energy, a new age of plenty made possible by this most "noble" form of energy, how fortunate it is that "none are hid from his heat."'[4]

Voicing the optimism of the 'Alternative Energy Movement' of the 1970s in *Rays of Hope*, Denis Hayes wrote:

The world has begun another great energy transition. In the past, such trans-formations have always produced far-reaching social change. For example, the substitution of coal for wood and wind in Europe accelerated and refashioned the industrial revolution. Later, the shift to petroleum altered the nature of travel, shrinking the planet and completely restructuring its cities. The com-ing energy transition [to the renewable path] can be counted upon to reshape tomorrow's world. Moreover, the quantity of energy available may, in the long run, prove much less important than where and how this energy is obtained.[5]

Earth Day was first celebrated in 1970. Hayes was one of its founders and also the chair of Earth Day 2000. Energy continues to be a main topic on Earth Day, which is now celebrated annually around the world, in nearly 150 countries, by millions of people. From the Earth Summit in Rio in 1992 to the G8 Summit in Tokyo in 2008 to the G20 Summit in London and the UN's Climate Conference in Copenhagen in 2009, and practically in all the major international conferences held during this period, energy has been a main topic. A growing concern about climate change has fuelled repeated discussions and debates over the need to establish goals for curbing CO_2 emissions and generating energy from renewable sources.

Around the world, conferences and forums on renewable energy are part of the annual routine. Many grassroots organizations have been formed around renewable energy issues. Many traditional organizations and institutions have adopted renewable energy programmes. Schools, colleges, universities and informal educational centres offer programmes and courses on renewable energy and other environmental subjects. A wide variety of educational and technological resources are now available to the public. An impressive variety of research has been conducted and documented. Books and articles have been published. Exciting projects utilizing a variety of renewable energy technologies have been successfully implemented. Unquestionably, there is a growing public awareness of and increase in public activity on renewable energy issues.

So, this much appreciation is well-deserved: these developments have indeed set in motion a transition to the renewable path. Going 'green', which almost always bears an implication for energy consumption, has gained a sudden, even dramatic, respectability within the last couple of years. But it is also obvious that the transition has not been fast enough to prevent us from getting more deeply entrenched in the nonrenewable path. The critical factor in our race to reverse the collapsing trend remains the urgency of action. So, we might ask, what would it take to inspire such actions to get over the barriers and to expedite such a fundamental transition? What would bring about that revolution?

Turning vision into action

Richard Buckminster Fuller said it well: 'You never change something by fighting the existing reality. To change something, build a new model that makes the exist-ing model obsolete.' Nothing will fuel that new model of reality, the vision of a sustainable and peaceable world, better than the revolutionary potential of renewable energy options.

In radical contrast to the nonrenewable energy systems, which are large, centralized and industry driven and controlled, the nature and scope of the renewable energy systems lie in diversity and choices. Most importantly, at its heart is the role of people, individually or collectively, to initiate, to choose, to act. Education is the key.

What follows, therefore, is not a centrally dictated or comprehensive list of things that need to be done. The incredible range of possibilities even defies such an attempt. Also, that would undermine the new and creative ways that people can act, individually and collectively, to move us all along a sustainable path. Rather, these are some inspiring examples and opportunities – just a glimpse – of concrete ways that individuals, communities, businesses, utilities, industries, educational and religious institutions, foundations, governments and non-governmental organizations (NGOs), from grassroots to international levels, can act, building up an infrastructure upon which the transition can flourish, upon which the vision of a sustainable and peaceful future can become a reality.

Conservation and efficiency

Who would argue against comfort and entertainment? After all, aren't these desirable aspects of what we would consider the 'good life' and a 'higher standard of living'? But it is a different story when the endless variety of consumer products which are supposed to comfort and entertain us begin to deplete Mother Earth's life-sustaining elements, bury her under their heaping wastes and poison her ecosphere on which we all depend. That is also such a waste of energy, across the production through disposal chain, a drain on the economy and, no less critical, a poisoning of our children's future. It is futile to talk about a transition to sustainability without emphasizing the necessity for conservation and efficiency, whether we are getting our energy from renewable or nonrenewable sources. Conservation, efficiency and renewables are fundamental to the transition to a sustainable future. They must go hand in hand, weighing the balance carefully so we can move beyond 'keeping the wolf away from the door for a bit longer' and actually transition to a renewable energy path.

There are many opportunities to conserve and use energy more efficiently in our daily life. Here are a few:

Use lighting, heating and cooling devices as selectively as possible. Open the shade or pull back the curtain before opting for the light switch. Shut off the light when it is not needed and turn off the TV when nobody's watching it. Paint the interior walls with light-reflecting colours, rather than light-absorbing colours, and save on wattage and the electric bill.

Add to this some of the energy-efficient products. 'Change a light bulb to save the world', along with a photo of a spiralling light bulb, have become a slogan and a symbol of going 'green'. Well, it may take more than that, but it's a good start. One 18-watt compact fluorescent energy efficient bulb gives out the equivalent light of a 75-watt incandescent bulb. It will last for 10,000 hours, compared to the 1000-hour life of the incandescent. It costs under $3. Some utility companies offer energy audits and rebates, and some environmental organizations sell efficient fixtures for a discounted price. According to Amory Lovins, a physicist and foremost advocate of renewable energy, replacing a 75-watt incandescent with a single 18-watt compact fluorescent over its life can save the electricity a typical US power plant would make from 350kg (770lb) of coal. As a result, about 907kg (2000lb) of CO_2, 9kg (20lb) of sulphur dioxide, which cause acid rain, and varying amounts of other pollutants would not be released into the atmosphere. Or, if electricity was being generated by nuclear power, it would cut down the production of half a curie of strontium-90 and cesium-137 (two high-level waste components) and about 25 milligrams of plutonium – about equivalent in explosive power to 386kg (850lb) of TNT, or in radioactivity, if uniformly distributed into the lungs, to about 2000 cancer causing doses.[6]

At some point, even the long-lasting bulbs will burn out. Make sure to dispose of them safely. On the whole, the energy-efficient bulbs are more advantageous than the conventional incandescent bulbs. But they contain mercury and must not be dumped as regular garbage. Some cities, states, municipalities, environmental organizations and stores – especially the ones selling them – have programmes or sites for collecting toxic wastes, which will accept burnt out energy-efficient bulbs. If there is not such a programme in your community, ask for one or initiate one.

European and Japanese appliances, such as refrigerators, dishwashers, stoves, washers and dryers, have been known for their efficiency and durability. US appliances have been known for energy guzzling and planned obsolescence. Some changes are transpiring, however. 'Energy Star' rated appliances are more efficient and cost-effective over the life of the appliances. Look for that label when you go appliance shopping.

If you need to carry drinking water, use reusable glass bottles. Plastic bottles take a lot of energy to manufacture and transport, contribute to global warming during bottle production and create waste through disposal. On 8 July 2007, an ABC News segment, 'Bottled Water, Wasted Energy?', announced that in the US alone, 30 billion empty plastic water bottles are thrown away annually – enough to circle the Earth 150 times – and 4 out of 5 of these end up in landfills. Some cities – Minneapolis, MN (proposed), Ann Arbor, MI, San Francisco, CA, and Salt Lake City, UT – are banning the use of public money to buy bottled water and some restaurants are serving tap water instead of bottled water to their customers. Most customers are comfortable with restaurants 'doing the right thing'. So, wherever you are in the world, demand that your municipalities or water departments supply you with clean water so you do not have to play into the hands of the highly profit-driven and mushrooming worldwide water-bottle industry. Tap water costs a fraction of the cost of bottled water. If necessary, you can add a filter to your tap for drinking water, both for use at home and carrying it with you in a glass bottle. There's also a growing concern that some chemicals, like Bisphenol A, commonly used to strengthen the plastic, may leach into the bottled water, then pass into the human body. So, rethink using plastic bottles.

Minimize packaging and buying packaged products. Not only does it take energy to manufacture packaging materials, polluting the environment at the same time, fancy packaging for many items costs more than the items themselves. And most of the fancy packaging ends up in the garbage in no time, anyway. Garbage consumes energy, money and other resources for its production and disposal, demanding landfills and incinerators. Wrap gifts by reusing wrapping paper or newspaper. If you're giving a book as a gift, and can deliver it directly, what wrapping is more attractive than the look of the book cover, anyway?

Looking at some of the food packaging, even of organic and natural foods (which I'm all for), I feel concerned that if the food doesn't kill us, the packaging will.

Whenever possible, buy locally grown fresh organic food – unpackaged, in bulk, in reusable bags or containers. It is good for your health and the soil. It minimizes transportation costs, energy consumption and greenhouse gas emissions. For the very same reasons, one may take a fresh look at the benefits of vegetarian and vegan diets over a meat diet – aggressively propagandized by the livestock industry, the latter is highly energy-intensive to produce, market, transport and, as more and more people are coming to realize, not good for one's health.

Buy toys without batteries, or buy solar toys. Millions of batteries are dumped every year from battery-guzzling toys. Use rechargeable batteries. Revive wind-up toys from their near extinction.

A well-insulated home or building conserves energy and saves money. Wear a sweater before upping the thermostat. Use fans instead of air conditioners. Open the window before turning on the fan. Whenever possible, share a home with more members and join a car pool. Support and ride on public transportation. If you need a car only occasionally, join an organization like 'Zipcar', an international membership-based car-sharing company providing automobile rental to its members, billable by the hour or day.[7] Ride a bike or simply walk. Such conservation measures are good economics as well as good for the health.

Think also about computers, monitors, video games, cell phones and TVs. As these electronic devices multiply in number and are discarded with greater frequency, only to be replaced by the latest models, many of them end up in landfills.[8] These contain mercury, lead, arsenic and others toxins, which contaminate land and water tables. In developing countries around the world, men, women, even children, with bare hands and no gloves or masks salvage whatever they can – plastics, metals, glass, microchips – from this hi-tech e-trash for pennies a day. Recycling policies and programmes are growing around the world, but their pace and number amount to a band-aid over a fast spreading cancerous wound. Illegal dumping is rampant. We have been so quickly drawn into – even addicted to (with the percentage growing among children) – these electronic devices, focusing on their benefits, that we tend to overlook the harmfulness of their toxic legacy. So, on the Reduce-Reuse-Recycle path, insist on the first. It'll conserve energy, protect human health, and heal the environment and the Earth.

The fuel-efficient car is yet another example. Amory and Hunter Lovins offered prophetic advice years ago: 'Improving America's 19-mile-per-gallon household vehicle fleet by three miles per gallon could replace US imports of oil from Iraq and Kuwait. Another nine miles per gallon would end the need for any oil from the Persian Gulf and, according to the Department of Energy, would cut the cost of driving to well below pre-crisis levels without sacrificing performance.'[9] Imagine what could have been avoided if Detroit took the advice, or if the lawmakers in Washington acted on it with some regulatory standards. The industry would probably be still moving on its own without the costly and dubious solution of an external stimulus.

Emissions from cars, buses and other road vehicles are the major contributors to both climate change and ground pollution, and constitute the most serious pollution factor around the world. Per gallon of gasoline burned, cars spew out 9kg (20lb) of carbon dioxide (CO_2) – the main contributor to climate change – into the atmosphere. The 1990s began with the transportation sector accounting for about one-third of the CO_2 emissions in the US and one-tenth of CO_2 emissions globally – nearly 1.7 billion tons. Vehicles also emit pollutants such as nitrogen oxides (NO_x) and reactive hydrocarbons (HCs) which cause smog in the lower atmosphere, or troposphere. Worsening this condition, the fuel-efficiency standards actually backslid during the 1990s for almost a decade while an increasing number of bigger gas-guzzlers (like the SUVs) kept infesting the roads. Around the world, the number of cars and other vehicles multiplied.

Thanks especially to skyrocketing gas prices in 2008, the demand for fuel-efficient cars began to grow. Hondas and Toyotas look more attractive each day. The demand for their electric-gas hybrid models, led by Toyota Prius (48mpg city, 45mpg highway, for the 2009 model, Environmental Protection Agency [EPA] estimate) and Honda Civic (40mpg city, 45mpg highway, for the 2009 model, EPA estimate), also featuring special pollution control measures, was at an all-time high until dipping slightly when the current recession hit. Hopefully, with economic recovery, fuel-efficient vehicles will continue to make a lasting impact on the direction of the auto industry as a whole. A growing number of other car-makers are putting out energy-efficient and hybrid models. Add to this trend the exciting prospect of electric cars charged by renewables.

All these measures are certainly positive steps toward fighting climate change and ground pollution. But we cannot stop there. As critical as climate change and ground pollution are, these are only two of the problems related to transportation. In a growing number of cities around the world, it is faster to go from one place to another by foot than by car – while, of course, nearly choking from exhaust fumes. According to a study released in September 2003 by the Texas Transportation Institute at Texas A&M University, Americans spend 3.5 billion hours a year stuck in traffic. People allocate more and more of their travel time to idling or sluggishly moving through traffic jams than driving. It is costly, too. The study found that 5.7 billion gallons of fuel are wasted in traffic jams annually, costing motorists an estimated $8 billion. Despite new road construction and mass transit alternatives, traffic jams are a daily experience for many.

However pseudo-romanticized by the barrage of car commercials, at an accelerated speed, the car ideology is driving people to dead-end roads. But this life-on-wheels does not merely strap people inside their cars, it also makes it increasingly difficult to get out of them: think about parking – the daily frustrating and costly urban hunting ritual for a little space in a car-infested landscape. We must address traffic jams before, literally, we get so stuck in them that it becomes nearly impossible to get out of them.

It is not just the cars; we must question the 'car culture' itself. Once meant to be a mere means of transportation, cars have become the Sacred Cows of the modern industrial society – no less venerable, and more worshipped than criticized. The current economic crisis has pushed the global demand for cars to an all-time low. Of course, we all hope that the economy recovers, and recovers quickly. The question is, will the car culture change if that happens? There is no guarantee. Human psychology teaches us that forced deprivation (caused by economic forces in this case) tends to actually fuel the attraction for the deprived object, and the consumption habits and values return quickly as the forces are lessened or lifted. Other forces are in the works. India's Tata Motors just launched its first fleet of the 'Nano', the world's cheapest car. Priced under US $2000, it is nicknamed the 'People's Car'. The demand for it has already far exceeded the first line of production. Tata Motors has on the drawing board similarly competitive models with slightly higher prices and higher emission standards for the European and US markets.[10] China is not sitting idle, either. The enormously costly auto industry bailout in the US is focused on making the car culture more palatable, instead of transforming it in a fundamental way through developing alternative options, such as public transportation, which would effectively minimize the need for cars and meet transportation needs in a socially, environmentally and economically responsible way.

Is life without (or with a minimal use of) cars possible? Among the few who offer a radical perspective on this issue are Ivan Illich, in his *Energy and Equity*,[11] and Jane Holtz Kay, in her *Asphalt Nation*.[12] The books provide an understanding of how the powerful auto industry systematically and ideologically fuelled our dependency on cars, which also led to co-opting public spaces, dominating the natural landscape with roads and parking lots, discouraging the natural human ability of walking and non-motorized biking, and sabotaging public transportation. It has done so at the public's expense, and to such excess that its own weight is now causing the industry to succumb under it. And the auto industry still has a powerful enough grip on the political administration – and the cultural psyche – to try to revive itself through even more public expense.

The car culture is spreading globally. The economic crisis can slow its pace, at least temporarily, but not transform it. What will transform it is our concerted effort at every level to promote education and awareness about this critical issue, re-evaluating the car culture, which can then lead to action at every level. The crisis, therefore, is also an opportunity. Lewis Mumford (1895–1990), the great American social philosopher and critic of reckless modernization, advised well: 'Restore human legs as a means of travel. Pedestrians rely on food for fuel and need no special parking facilities.'[13]

How about trying living in the dark? Or at least, with essential lights only? On Saturday, 20 October 2007, San Francisco looked darker than usual from 8:00 to 9:00 p.m. Lights Out San Francisco, a community-based grassroots initiative, organized the educational and conservation event to voluntarily turn off all nonessential lights for the hour. The participants included such landmarks and buildings like the Golden Gate Bridge, Transamerica Pyramid, City Hall, Coit Tower, Bay Bridge, Palace of Fine Arts and many other institutions, businesses and residents. Nate Tyler, executive director of Lights Out San Francisco, got the idea from Earth Hour in Australia. On 31 March 2007, Sydney went dark for an hour from 7:30 to 8:30 p.m., with participation from over 2.2 million people. This saved 10.2 per cent of energy, representing a reduction of 24.86 tons of CO_2, equivalent to taking 48,613 cars off the road for an hour. Lights Out San Francisco estimates an energy saving of 15 per cent. The organizers also gave away 100,000 compact fluorescent light bulbs, donated by Pacific Gas & Electric Co., so the energy savings and carbon reduction would continue past the event.

Encouraged by the success, the organizers have taken the initiative to a global level. On Saturday, 28 March 2009, at 8:30 p.m. local time in nearly 4000 cities in over 80 countries around the world – Australia, Brazil, Canada, Chile, China, France, Greece, Hong Kong, India, Indonesia, Japan, New Zealand, Peru, Rome, Russia, Sweden, Switzerland, Thailand, the UK, the US, Vietnam and others – almost a billion people were mobilized to observe 'Earth Hour 2009' by turning off lights in homes, office buildings and landmarks for one hour. The campaign slogan 'Fight climate change with the flip of a switch!' is catching on. Its message is for year-round conservation of energy. The invitation to participate is open.[14]

We can start by doing simple things in our everyday life. Look for books like *50 Simple Things You Can Do To Save The Earth*,[15] *50 Simple Things Kids Can Do To Save The Earth*,[16] *Save Our Planet: 750 Everyday Ways You Can Help Clean Up the Earth*,[17] *Design for a Livable Planet*,[18] *Consumer's Guide to Effective Environmental Choices*[19] and *The Green Book: The Everyday Guide to Saving the Planet One Simple Step at a Time*.[20] These books speak to us today more than ever before. We can always put together our own list of actions.

So, much can be achieved through conservation and efficiency. In myriad ways, in our daily lives each of us can act toward these goals. They add up. Conservation and efficiency not only cut down on depleting nonrenewables, they also reduce the size requirements and expense of renewable energy technologies. Dan Reicher, Director of Climate Change and Energy Initiatives at Google.org, in his testimony before the Congressional Joint Economic Committee on 30 July 2008 cited a McKinsey and Company study, saying, 'that additional investments of $170 billion annually for the next thirteen years would be sufficient to capture the energy productivity opportunity identified in the 2007 report – i.e. cutting projected global demand by 2020 by at least half'. It calls for a global policy and every effort must be made to achieve it.

An initiative can be taken, and the benefit can be felt, at the household level. A conservation-conscious home with efficient light fixtures and appliances will need a smaller PV or other renewable energy system to meet its electricity needs, saving money on the utility bill as well. Such achievable individual steps, adding up to a massive scale and fuelling the power of the people, will influence the political will – to make policies and act on them – building a bridge to the transition. Let's take this household equation to community, national and global levels. And to conservation and efficiency, let's now add a third required element to expedite the transition: renewable energy education for action.

Renewable energy education for action

On the possibility of transition to the renewable energy path, the best news under the Sun are the many ingenious breakthroughs – both in developments and applications – the renewable energy technologies have experienced over the years. But so entrenched have we become in the nonrenewable path, and so unfavourable has the politics, and thereby the mainstream media, been to these developments, that public knowledge of the down-to-earth revolutionary possibilities of renewable energy technologies has been severely lagging. Education is the bridge. Concerns over skyrocketing fuel prices, climate change and energy wars are certainly prompting us to build the bridge. What we need to do, therefore, is to expedite the educational process with the utmost urgency, around the world, through every possible channel, and into actions. The goal is a powerful synthesis of education and action. Nothing is more frustrating than being aware of something but not being able to act on it. Powerlessness breeds denial. Turning it around – nothing beats the power and potency of mass education which is intentionally geared for action in bringing about a social transformation. The nature of renewable energy technologies is inherently attuned to that kind of synthesis of education and action.

Empower people

One of the revolutionary features of renewable energy technologies is that many of the systems can be designed and implemented both on personal and community scales. Anywhere in the world one can set up a personal little electricity generating station with photovoltaics or a wind turbine or a micro hydroelectric system, depending on the available power source. One cannot do that with nuclear power or fossil fuel-powered plants. Just the same, a community can choose to set up a larger renewable energy power plant for the entire community. Whether in urban, suburban or rural settings, a community can implement renewable energy projects with a variety of systems and sizes, matching the need and the nature of the community. These can be 'stand-alone' systems, independent and completely under the user's control, or 'grid-connected' hybrid systems, connected to the power grid. For whatever amount of electricity is generated through a grid-connected system, the owner is also the generator and seller of energy to the grid. Now the owner's utility bill is calculated by subtracting the amount of energy generated through the system from the total amount of energy used by the owner. The transaction is called 'net metering'. In effect, a grid-connected system runs the owner's meter backwards for the amount of energy generated. Currently, at least 36 states in the US permit net metering. Because the electricity is generated at decentralized sites, the process is called 'distributed generation'. It adds capacity to the utility grid. Blackouts take notice! With an increasingly scarce and insecure power supply from conventional sources, distributed generation through renewable sources is a secure advantage which both the owners and utilities can count on. That is a radical departure from the conventional ways of generating and distributing energy from a centralized and utility-controlled power plant. Don't underestimate what this radical transformation of power – however small in scale – can add up to. Every candle can light another candle. Likewise, let each of the small solar electric systems inspire another solar electric system. . .and another. . .and another. Imagine the possibilities!

People pay for what they value

Knowing what the options are, and what each option means, some people even choose to pay a little extra for their 'green power'. Just as solar options were becoming available to customers, years ago Jonathan and Shirley Robinson chose to pay an extra $4 per month for the grid-connected PV system installed on their roof as part of the PV pioneer programme run by Sacramento Municipal Utility District, California (SMUD).[21]

In Michigan, customers voluntarily chose to pay extra to share the cost of the electricity generated by a 600kW wind turbine erected by Traverse City Light and Power, a municipal utility.[22]

In 2008, Boston's Logan International Airport installed 20 utility-connected wind turbines, 1kW each, to meet the equivalent of 2 to 3 per cent of the energy needs of the airport's Logan Office Center. It cost $140,000. With a projected annual saving of $12,000–15,000 on Logan's energy bill and a ten-year payback period, it is a wise economic decision – especially as the system will have another ten-year warranted life, powered by free fuel. Still, the project's main emphasis remains on its environmental values. Sam Sleiman, director of capital programmes and environmental affairs of the Massport Authority, which runs the airport, says, 'We're not doing it just for the payback. We're doing it for the environmental benefit and the energy it's going to produce.'[23]

It also reminds me of a conversation I had a few years ago with a rural cobbler in Bangladesh who chose to install a small stand-alone PV system on his humble two-room mud-thatch hut. His workplace was a mat at the foot of a tree in the village's open-air marketplace which did not need artificial lighting and his lighting needs at home were minimal, adequately met through two kerosene lamps. Still he decided to buy the PV system, made affordable through microcredit financing. When I asked him about it, he answered that he was of course very pleased with the better lighting, odour-free too, that the PV system provided. But most importantly, he said: 'I'm a poor man. What makes me most happy is that the solar lamp helps the environment and that I'm now able to do something good for my children's future.' This is what we refer to these days as leaving a smaller carbon footprint and caring for our children's future.

These examples tell us something about what we do everyday when we go shopping for any number of things: a pair of shoes, a piece of furniture, a car, a home. We are willing to pay for what we value, which can mean paying a little extra compared to what we are used to paying, if we have the choice. Especially at this stage, the success of the transition to the renewable energy path will have to be mainly value driven. It certainly helps that the cost of renewable energy technologies is on a downward trend, looking increasingly attractive against the conventional fuel bill. But to accelerate the process, we need more. The good news is, like the Robinsons, the customers of wind energy at Traverse City, Sam Sleiman of the Massport Authority and the cobbler in Bangladesh, gradually people all over the world are beginning to see, from a holistic perspective, the value of renewable energy beyond the price tag. To them, clean energy, a healthy environment, a peaceful world, and a sustainable future which our children and their children can look forward to are important, too. And it is worth subsidizing the transition. In doing so we are also laying a new foundation for society – a global society. We must build upon that foundation, urgently.

Exposure educates, demonstration motivates, success replicates success

Very few areas are as laden with scepticism, unfair discredits and misleading propaganda by Big Oil and the nuclear industry as the area of renewable energy. A lot of people still consider renewable energy technology to be an exotic fantasy of the future, worry that lights will not come on when the Sun goes down, and, certainly, that it is never going to be sufficient to meet our energy needs. You cannot blame them. Big Oil and the nuclear industry spend billions of dollars to fuel the propaganda. On the other hand, no matter what others say, nothing is more convincing than seeing an example of something that actually works. So, demonstrate successful projects. Be accurate, informative and uplifting, just like any effective campaign. Document, evaluate and explain the steps, opportunities, advantages and barriers these projects face both commonly and uniquely. Keep it simple, comprehensible and human. Praise initiative, persistence and vision.

'Solar Home Tours' began some years ago with only a few people opening up to the public their renewable energy-powered homes – mostly by PV and solar hot water systems. Repeated over the years, they have been gaining momentum, becoming national annual events across the US. More and more homes open up and more and more people join the tours. Along with the demonstration of solar energy systems, all the fascinating human stories behind the personal and community-based initiatives that come to life during the tours are simply fascinating and inspiring. It brings to light the massive potential power of small-scale initiatives and actions. If one were to investigate into large-scale projects that are happening around the US, even around the world, it is amazing how one would see that the impetus behind some of those projects is rooted in small-scale efforts – seeds from which they sprouted and blossomed. They are more than case studies; they are human stories of vision and action, as well.

Let's look at the story behind Gardner, the 'first solar electric neighborhood in the world'. Gardner, a town about 80km west of Boston, Massachusetts, has long been known as 'The Furniture Capital of New England'. By the 1980s, the businesses were not doing so well, the factories were closing or moving away, and the town economy dwindled. For many long-time residents, it was hard just to hold on to their homes. It was a perfect time for someone to reach out to the town with a proposal that would help the town's economic survival. The opportunity caught the imagination of Joan Bok, the chairperson of New England Electric, the town's utility provider. She teamed up with Steven Strong, an engineer and president of Solar Design Associates, who would design and coordinate the project. In addition to providing free electricity to the participants, the project would also conduct research on the reliability, variation in production of electricity during the year and effect that a cluster of PV installations has on a single power distribution line. The findings would also provide some invaluable understanding about the scope of grid-connected systems, in general.

Steven Strong himself is an inspiring pioneer-activist in the renewable energy field. While serving as an engineering consultant on the Alaskan pipeline during the early 1970s he came to the realization that 'going to the ends of the earth to find fossil fuel was only an interim measure, not the answer to our long-term energy needs'. He returned from the project transformed by the experience and founded Solar Design Associates, a design firm specializing in sustainable energy options, primarily photovoltaic systems. Over the last 30 years, Solar Design Associates has evolved into North America's oldest, largest and most respected building design firm dedicated to the artistic integration of solar energy in homes and buildings.[24] Steven Strong's book, *The Solar Electric House*, first published in 1987, remains one of the best design manuals for home-scale photovoltaic systems on the market.[25]

A Gardner proposal was presented to the town's people who embraced the project that would, through net-metering, reduce their electricity bill according to the amount of electricity generated by the PV systems on their individual roofs. By 1985, 2kW to 5kW grid-connected PV systems were installed on 30 homes, the Town Hall, the library, a Burger King restaurant, a furniture outlet and a community college. The Gardner solar electric neighbourhood was born. Research followed, including one study supported by the Electric Power Research Institute (EPRI), headquartered in Palo Alto, California. EPRI was interested in the results of the Gardner project for possible applications throughout the nation. It sponsored research by Ascension Technology, Inc. and New England Power Service Company, both of Massachusetts. Presented at the 1989 Power Distribution Conference at the University of Texas at Austin on 23 November 1989, the report declared the experiment a success.

Prompted by a lot of media coverage, especially during the early years of the project, visitors flocked to it from around the world. I have had the pleasure of going on a Solar Homes Tour to Gardner led by Steven Strong himself. Over the years, I have gone there many times with my friends, family, students and visitors from around the world, meeting and talking with the residents. Visitors from Japan and Germany – the world's top two producers of photovoltaic electricity today – visited Gardner and were amazed by what they saw during the years when photovoltaic technology was practically unknown in these countries. The 'first solar electric neighborhood in the world' in Gardner is a success story, rooted in a vision, personal initiative, care, action and persistence which has inspired many such projects both within the country and around the world.

'Going Solar', a project of the Interstate Renewable Energy Council (IREC), founded in 1982 as a nonprofit organization and headed by Jane Weissman, a pioneer renewable energy policy expert, is an excellent audio-visual educational resource kit that demonstrates – through examples of homes, schools and businesses – how to build communities with solar energy. The IREC online newsletter itself is an excellent source of information on new developments in the renewable energy field.[26]

'Solar Boston' was formed in 1999 as a partnership of community organizations and solar energy companies to promote the use of renewable energy technologies in the Greater Boston area by providing information and assistance to interested consumers. It was initiated in response to a call for action by a few of us following a conference we were attending on how to bring solar energy to communities. The conference, 'Communities Going Solar', was sponsored by IREC and the Massachusetts Division of the US Department of Energy. The attendees came from all over the country. The remarkably diverse Solar Boston partnership growing out of it included Boston Area Solar Energy Association, Dudley Street Neighborhood Initiative, Episcopal Power and Light, Fenway Community Development Corporation, Heliotronics, Inc., Interstate Renewable Energy Council, Massachusetts Energy Consumers Alliance, Solar Works, Inc., Tufts Climate Initiative, Wainright Bank and Trust Company, and a solar consulting firm, Zapotec Energy.

The Solar Boston Steering Committee, made up of volunteer representatives from the partnership organizations, met regularly for four years to plan and offer services that included community outreach and education, consultation, technical training, market development, project planning, financing and policy development. It also included administering project funding by the Massachusetts Technology Collaborative (MTC), the state agency which supports renewable energy installations across Massachusetts with funds from the Massachusetts Renewable Energy Trust. The committee was assisted by minimal staffing funded by the Department of Energy's Million Solar Roofs Initiative (MSRI), of which Solar Boston served as a regional partner. As a result, the Greater Boston communities now have numerous schools and homes with PV systems, mostly 2kW grid-connected systems, lighting the way for more such installations. Also, Solar Boston conducted many workshops, lectures and other events in a wide range of settings to educate the public about solar energy. A Greater Boston solar energy infrastructure was established.

The funding for staff assistance ended in 2004, but a great achievement of the partnership was that by that time much of its functions had been internalized by one of the partners, Massachusetts Energy Consumers Alliance (Mass Energy), which also generously hosted Solar Boston over the years. A nonprofit oil consumers' cooperative at the time of joining the Solar Boston partnership, Mass Energy has evolved, under the leadership of its executive director, Larry Chretien, into a major force behind renewable energy development in Massachusetts, offering an expanded range of programmes and services.[27]

Additional good news is that Solar Boston was approached by the City of Boston to lend its name and logo to the Solar City Strategic Partnership role in the Solar America Initiative (SAI) Market Transformation. Solar America Initiative is the refashioned name of the US Department of Energy's Million Solar Roofs Initiative, aimed at promoting renewable energy technologies, mainly photovoltaics and solar thermal systems, across the country. The City of Boston's proposal for the SAI Partnership – as Solar Boston – has already gained support from MTC, utility provider Keyspan, an anonymous foundation, IBEW Local 103 electrical workers scheme and others. With the supportive network still growing, Solar Boston will have an unprecedented opportunity to promote renewables throughout the city. The City's long-term goal for generating solar electricity is to increase the amount of installed solar capacity from 0.5MW today to 25MW by 2015. The mayor of Boston, Thomas Menino, says he would like to turn 'Beantown into Greentown'.[28] I'm all for it!

The Solar Fenway Demonstration Program of the Solar Fenway Committee was launched to educate people about renewable energy as well as to install some solar systems in the Fenway, a Boston community with 40,000 people in 3.9km². The volunteer committee of community residents has been promoting renewable energy through forums and workshops since 2002. In response, and in collaboration with Solar Boston and the Fenway Community Development Corporation, a nonprofit community agency, the demonstration programme was conceived and subsequently launched. The first installation took place in December 2005. It is a 2.4kW grid-connected PV system installed at the Boston Arts Academy, a pilot high school. The system generates about 2860kW hours of electricity for the school annually, lowering the energy bill by that amount, and offsets about 1590kg (3500lb) of CO_2 emissions compared to electricity produced at coal-fired plants, a major cause of global warming. The $26,000 system was jointly funded by the MTC, the Mission Hill-Fenway Neighborhood Trust, the City of Boston, Boston Public Schools and Solar Boston/Mass Energy Consumers Alliance. A mayoral inauguration took place on 3 April 2006. It brought together community activists, state and city officials, students, teachers and administrators, funders, installers and journalists in a grand celebration of collaboration and action. And it is just the beginning. Students at the school have continued to use this as an educational tool and, as part of their 'Boston Arts Academy Alternative Energy Project', have prepared and presented to the City of Boston a plan to implement more renewables across the city!

In March 2007 the second installation of the Solar Fenway Demonstration Program, a vertically mounted 1.5kW grid-connected PV system, took place at Fenway Views, a condominium building with some affordable housing units owned by the Fenway Community Development Corporation (Fenway CDC). The system generates about 1790kW hours of electricity annually, reduces the building's energy bill by that amount, and offsets about 1000kg (2200lb) of CO_2 emissions. The $17,000 system was jointly funded by the Massachusetts Technology Collaborative, Mission Hill-Fenway Neighborhood Trust, and the Fenway CDC. A June inauguration followed by a keynote speech by Pulitzer journalist and author, Ross Gelbspan.

The Fenway also prides itself on some of the other renewable energy systems in the community. Northeastern University's Curry Student Center has a 26kW grid-connected PV system. Installed in 1994, it is one of the pioneering initiatives in Massachusetts and a leading solar installation by a university in the country. While reducing the university's electricity consumption from nonrenewables, the 90-module system also offsets 13.5 tons of CO_2 emissions into the air.[29] Genasun, a West Fenway company offering energy system design services and distributing off-grid PV systems, marine batteries and charge controllers, has a small PV module installed above the entrance to its office/warehouse to generate electricity to test batteries.[30] In 2008, the Boston Red Sox's Fenway Park stadium installed a solar hot water system which supplies 37 per cent of its annual hot water needs and offsets 18 tons of CO_2 emissions.[31]

A major leap in solar installations in the Fenway is in the works. The Fenway Center, a huge $500 million complex of 330 apartments, 34,400m² of office space, 8,400m² of stores and a 1290-space parking garage, situated over the turnpike by the side of Fenway Park, is scheduled to begin the first phase of its development in the summer of 2010. It will include a 650kW, 1200-module solar array PV system, generating electricity equivalent to all the power requirements of the Yawkey commuter rail station and an additional 100 apartments located in the complex. Once completed, it will be the largest solar facility in Massachusetts. John Rosenthal, President of Meredith Management Corporation, the owner-developer of the complex, estimates that tax credits will allow him to recoup the solar installation costs of $7.4 million in four years, followed by paying off the hardware costs with the proceeds from the sale of electricity generated by the system in subsequent years. Rosenthal's decision to take on this project is rooted in his values as a long-time environmental activist who was jailed three times in the late 1970s and 1980s for protesting against nuclear power plants. 'To leverage my business to produce green power is a dream come true for me,' he said to Casey Ross, a *Boston Globe* writer. In the past Rosenthal founded two nonprofit corporations, Friends of Boston's Homeless and Stop Handgun Violence. To promote the use of renewable energy he has recently launched his own power company, Here Comes the Sun LLC.[32]

The Solar Fenway Demonstration Program is having a positive effect. Interest in solar is growing among community residents, business-owners, developers, and a variety of educational and medical institutions which characterize the Fenway. The success of the programme has also united Solar Fenway and the Fenway CDC in a stronger partnership, with Solar Fenway now acting as a committee of the Fenway CDC to advocate for solar projects in the community. The Solar Fenway committee now also includes representatives from Husky Energy Action Team (HEAT). HEAT is a student-organization at Northeastern University, committed to reducing carbon footprints throughout the campus, including integrating renewable energy technologies in the university buildings.[33]

Solar Fenway has initiated contacts with co-op and condominium owners, Northeastern University, the Museum of Fine Arts, the Boston Red Sox, Boston Latin School, McKinley School, developers and businesses in the Fenway, proposing to integrate more solar into the community. The organization will continue to contact others and looks forward to collaborating with the city, state and national initiatives. The demonstration programme has taught some invaluable lessons. By building on that experience, to accelerate more solar installations throughout the community, Solar Fenway is gearing up to launch a Solar Community Pioneer Project, the first solar community of Boston. The Solar Fenway brochure states as its mission, 'Turning the Vision of a Sustainable Community into Action!'[34] The revolutionary nature of renewable energy technology offers such options and power to the people. People just have to act on it.

It inspires me to look at the 46W stand-alone PV system we installed in our home in 1986. Sitting on the south-facing window sill of our fifth floor condominium unit in the Fenway, with the battery box placed inside and under the window, it has been supplying electricity for a room with two 15W fluorescent lights, a table lamp, a small table fan and a record player diligently and reliably, around the year, for over 20 years! All I had to do was to replace the set of two interconnected 6V, deep-cycle batteries twice. The room is also equipped with a variety of solar cookers – both home-made and factory-made – well-used over the years. The PV system also has the capacity to power our 'Tulsi Hybrid' solar cooker that can cook three ways, day and night, year-round: by direct sunlight, being plugged into the regular household current (110VAC), or by solar electricity from the PV system (12VDC converted to 110VAC through an inverter). Last but not least, the battery in our digital camera, too, gets recharged by the PV system.

Through all the tinkering and experiments I have conducted on this system, it also humbles me to think how much this basic little system has taught me. It has provided me with a prototype, as I have gone on to work on solar projects both in our own community and around the world, resulting in thousands of installations for all kinds of purposes. And it further confirms the power of demonstration to think how the same system has inspired so many visitors, students and activists from around the world who have passed through this room then gone on to pursue their solar mission.

I do not know if there is any other home in the world which has so successfully demonstrated the value and practicality of solar living than the home of Bill and Debbi Lord in Kennebunkport, Maine. Designed by none other than Steven Strong, the 256m² residence with two bedrooms, two-and-a-half baths, living room, kitchen, entertainment room and computer loft integrates a range of sustainable features. Covering the south facing roof, it has a 4.2kW, 36m² grid-connected PV system and a 45m² solar thermal collector to heat water, to be stored in two 500-gallon tanks inside the house. For 15 years the Lord's home, despite Maine's harsh and cold winters and modest New England solar resources, has produced its own electricity, heat and hot water from the Sun. Year after year they have generously opened their home for Solar Home Tours and throughout the year welcomed visitors of all ages from around the world. It has been featured in countless magazines, television shows and documentaries. The 'Maine Solar House', as it has come to be known, even has its own website. On 1 February 2010, its visitor count was 1,376,452, and counting![35]

In Chicago, people can tour the city's museums not only for what is exhibited inside them, but also what sits on them. As part of the Solar Museum project, seven Chicago museums – the Art Institute of Chicago, the Museum of Science and Industry, the Mexican Fine Arts Center Museum, the DuSable Museum of African American History, the Chicago Historical Society, the Peggy Notebaert Nature Museum and the Field Museum – have installed grid-connected PV systems on their roofs.[36] To live their mission of preserving and exhibiting art and other objects indefinitely museums will continue to need energy. There is no better guarantor of that than the Sun!

Businesses are going solar. BJ's Wholesale Club, a massive consumer goods chain, has installed utility-interactive PV systems in three of its locations in Massachusetts, Rhode Island and Pennsylvania, generating a total of 108kW of electricity. Walmart stores and Applebee's restaurants have installed PV systems at some of their branches. Whole Foods Market, the world's leading natural and organic foods supermarket, which started off by installing PV systems in some of its stores in sunny California, now has a PV system at its distribution centre in Cheshire, Connecticut. The project partners, Whole Foods Market, the Connecticut Clean Energy Fund (CCEF), created by the Connecticut General Assembly to promote development and commercialization of clean energy technologies, and SunEdison, which consults, installs, finances, delivers and manages solar-generated electricity with no upfront capital costs at or below current retail rates to commercial, governmental and institutional grid-connected clients, held a dedication ceremony on 5 September 2006, unveiling the 121kW rooftop PV system that will supply 10 per cent of the facility's total energy annually. In addition, the system will offset 65 tons of greenhouse gas emissions each year. On 16 January 2007, New England's largest PV installation was unveiled at the office product company Staple's distribution centre in Killingly, Connecticut. Built in partnership with the CCEF and SunEdison, the 433kW system covers 6875m² of roof space of the 27,870m² distribution centre and has the capacity to produce enough energy to meet 14 per cent of the centre's energy needs. In October 2006 Google announced it will install a 1.6MW PV system at its Mountain View headquarters near San Francisco. Scheduled for completion during the spring of 2007, the system will generate about 30 per cent of the headquarters' power. Apart from meeting its own energy needs, Google is deliberately making it a demonstration project. 'We hope corporate America is paying attention. We want to see a lot of copycats of this project,' said David Radcliffe, the company's vice president of real estate, in *USA Today*.[37]

The George Robert White Environmental Conservation Center of the Boston Nature Center, a sanctuary and urban environmental education centre of the Massachusetts Audubon Society, is a model of environmental sustainability and energy efficiency. The building includes PV roof shingles, pole-mounted PV arrays, PV-powered street lights, a geothermal heat pump, a solar hot water system, daylighting and energy-efficient features. The centre has also installed a Heliotronics data-monitoring system with an interactive touch-screen kiosk to educate the public about renewable energy technologies and efficiency.[38]

Hull, a town 30km southwest of Boston Harbour, is the home of the Hull Wind 1 and Hull Wind 2 turbines. It is a success story that is blowing in the wind across the coastal towns of Massachusetts. Hull Wind 1, a 660kW turbine, has been operating here successfully since its installation in December 2001, generating 1,500,000kW hours of electricity annually, enough to power 250 homes. The turnkey project, with a life of 20 years or more and costing $700,000, was funded with accrued taxpayer monies. It is installed at a site adjacent to Hull High School, where the townspeople had installed a 40kW turbine in 1984. During the lifespan of this earlier turbine, until 1996, the Town of Hull saved $70,000. Operating superbly at this site, jutting out into Boston Harbor and also referred to as 'Windmill Point', Hull Wind 1 generated over 11.5 million kilowatt hours of electricity for the town by May 2009. Building on its success story, in June 2006, Hull Municipal Light inaugurated Hull Wind 2, a 1.8MW turbine which can generate enough power for 750 homes, offsetting 3000 tons of CO_2 a year. In May 2009, Hull Wind 2 hit the 11 million kilowatt hours mark of power generation. It has all been possible because a few townspeople – including Malcolm Brown, a retired philosophy professor and Commissioner at Hull Municipal Light, John McLeod, Operations Manager of Hull Municipal Light, Andrew Stern, a hi-tech expert and consultant, and Anne Marcks, a teacher at Hull High School – voluntarily united as Citizens for Alternative Renewable Energy (CARE) to initiate the Hull Wind project in collaboration with Hull Municipal Light.

Hull Wind tours are increasingly popular among visitors of all ages from near and far. Standing under one of these tall turbines silently doing its job way up in the air without interfering with whatever is going on around its base of just a few feet in diameter on the ground below, or watching the turbines from a boat that takes you around them as part of a tour, is an unforgettable experience. And the Hull

Wind website – thanks especially to Andrew Stern and other volunteers – is updated diligently and frequently, documenting the total kilowatt hours as the elegantly spinning turbine blades keep cranking up the numbers.[39] At the 5,000,000kW hours of Hull Wind 1 celebration on 5 March 2005, Malcolm Brown announced, '5 million is good, 10 million is better!' Within a little over a year's time, Hull Wind 2 was inaugurated. CARE is already planning its next set of turbines and inspiring a vision of a 'string of pearls' of wind turbines along the coastal towns and beyond! Four more turbines are in the works. Not to be ignored, on 31 January 2007, Hull Wind also installed a smaller 1.8kW 'Skystream' turbine, made by Southwest Windpower and costing $12,000, which will generate enough electricity to power the town's Weir River Estuary Center, where visitors can learn about clean energy at the foot of the turbine. As the 'I Visited the Windmill in Hull, Mass' button says: 'HULL-E-LUIA!'

The American and European Tour de Sol solar and electric car races, held annually, are showing that there are alternative ways to power transportation. One of their younger counterparts, the Junior Solar Sprint, is jointly sponsored by the New England Sustainable Energy Association (NESEA), Boston Area Solar Energy Association (BASEA) and MIT. Held annually on MIT's running track in Cambridge, it brings together students from Greater Boston middle schools to show and race their model PV-powered cars. There were about 100 of those cars in 2008! Sometimes there is a lifesize solar electric car, such as a 'Solectria', or a fully functioning MIT PV-powered car – with PV cells covering its body – parked or driven around the track to demonstrate that these fledgling models do have a future!

PV and solar thermal systems are demonstrating their appropriateness even in countries that receive much less sunlight than most other regions in the world. They are gaining momentum as supplementary sources in the energy mix, powering roadside or railroad signals, individual homes and even entire communities. In the UK, better known for London fog than sunlight, Beddington Zero Energy Development (BedZED) shines brightly. It is the UK's largest eco-friendly community. Built in 2002 in Beddington, in the London borough of Sutton, it is a community of 100 homes, plus shared facilities and workspace for 100 people. With a range of ecological features, it is a model of sustainable living. Along with several other energy conservation measures, BedZEd's use of passive solar technology for heating and photovoltaics for electricity speaks loudly for its goal to 'create a "net-zero fossil energy development", one that will produce at least as much energy from renewable sources as it consumes'. BedZED is a small community that is open to the world. Visitors from all over the world are welcomed there to join one of the regularly scheduled tours to experience what the organizers call 'one planet living'.[40]

Vauban in Germany is called an 'Energy-Surplus' neighbourhood. Located 4km south of Freiburg, Germany's 'greenest city', and spread over 340,000m², this carbon-neutral community of 5000 people with privately owned multifamily residences, co-ops, co-housing, affordable housing and a community centre is built on a former French military base converted during 1990–2006. It features PV for electricity, solar collectors for hot water, solar heating and cooling, a combined heat and power station burning wood chips, recycling, organic gardening and food production, biodigester, public transportation, bikes, pedestrian paths and shared spaces. And it is 'car-free' in the sense that ownership of cars by anyone, and for any purpose, is strictly limited, and so is allowing outside vehicles to enter the neighbourhood. With roofs covered with PV arrays, 'People make more money by selling electricity to the grid than they pay for heat and hot water and electricity', producing more energy than they ever use![41]

The solar catamaran 'sun21', designed by MW-Line Switzerland, is a 14m boat completely powered by a 65m^2 photovoltaic array, also serving as its roof. The boat left Seville, Spain, in November 2006, on a 11,300km journey to arrive in New York on 8 May 2007, to be greeted by a cheering crowd awed by the fantastic feat. Sailing at the speed of 5 to 6 knots per hour, about the same as most yachts, causing no noise, air or water pollution, it set the world record for transatlantic crossing by a completely solar-powered boat. The catamaran 'sun21' is now in the *Guinness Book of World Records*. There were five crew members: Mark Wüst, Michel Thonney, Beat von Scarpatetti, David Senn and Martin Vosseler, all from Switzerland and all deeply committed to working toward a more sustainable future. The purpose of the courageous journey was to demonstrate to the world the great potential of renewable energy. That they did![42]

Martin Vosseler, 60, has continued his journey to promote solar energy in other ways. A physician and the founder of the Physicians for Social Responsibility in Switzerland, he devotes much of his time to this mission which he considers his utmost priority. On 15 August 2008, shortly before noon, a group of us – friends, officials from the Swiss consulate, curious pedestrians and a couple of journalists – gathered at the intersection of the Boston Common and Commonwealth Avenue to greet Martin, scheduled to arrive there at noon to complete a walk he started in Los Angeles on 1 January. And he did arrive, Swiss time! With him was Claudio Beretta, also Swiss and an undergraduate in environmental studies at the Eidgenössische Technische Hochschule (ETH) in Zürich, who joined Martin in Newark for the rest of the walk to Boston. Martin was already inspiring the next generation! Wearing a long sleeve blue T-shirt – boldly marked 'SunWalk2008.com' (now a discontinued domain), a pair of khaki shorts, rugged sandals, and wheeling a duffle bag containing all his belongings, he just completed a 3,333-mile (5364km), 6,666,666 step walk through California, Arizona, New Mexico, Texas, Oklahoma, Arkansas, Tennessee, Virginia, Washington, DC, Maryland, Pennsylvania, New Jersey, New York City, Connecticut, and finally Massachusetts. Braving all kinds of weather, alongside the highways running through dramatically changing landscapes, he kept on walking for seven-and-a-half months. He made stops along the way to meet with local environmental activists, students, teachers, artists, public officials, business-owners, writers, interviewers, journalists – anyone who would like to host him for a night or two or just listen to the message he is committed to spread: the Earth is running out of nonrenewables, we are destroying the environment, poisoning ourselves, turning to violence against each other. It is a dead-end road. But there's hope. The solution, Martin believes, lies with solar energy. 'There's enough Sun for all of us,' he says, as he walks his talk.[43]

So, promote renewable energy any way you can. Take personal action. Use the media: newspapers, magazines, posters, bumper stickers, billboards (lighted by PV!), cereal boxes, T-shirts, TV, radio and internet. Power public events like Earth Day,

Boston Solar Day and energy fairs with renewable energy. Organize more of these kinds of events and, just as important, initiate activities where they are not happening. Organize solar tours. Implement demonstration projects in your community. Show that renewable energy technologies are not just exotic technologies of the future. They are here – to stay – successfully. Let's build on that success!!

Affordability motivates action

Californians have led the way on the renewable path in the US. They seem to have taken the challenge even more seriously since the energy crisis in 2001 which forced the state's economy to virtual bankruptcy, and caused countless stories of inconvenience and suffering. The result is showing in the surging popularity of renewables since the crisis. In a survey conducted by Maryland Marketing Research, Inc. for the California Energy Commission (CEC) Renewable Energy Program, more than 50 per cent of homeowners are willing to pay more for a home already equipped with PV or wind technology. The survey also found more than 60 per cent of homeowners would be more interested in a renewable-technology-fitted home versus a home that is not. The blackout of August 2003 added more fuel to this motivation for renewables.

However, when it comes to actually paying for renewable technology systems, there is a real challenge for most homeowners. The challenge translates into having to pay for years of electricity or heat all at once. That is a barrier even if the system is cost-effective over time. The problem is similar to the one faced by an average consumer shopping for a car, a refrigerator or a computer. If the customer had to make the full payment right away out-of-pocket, chances are that such a purchase would be prohibitive. So here comes the credit card or a bank loan – one of the financing options which makes such conveniences affordable. Some banks have pioneered special incentives toward renewable energy financing. GMAC (General Motors Acceptance Corporation) began offering mortgage financing for solar homes – renewable energy technology integrated homes – at a rate which is at or below the commercial rate. Wainwright Bank in Massachusetts offers loans at a lower than commercial interest rate for renewable-energy-related expenses. Some banks offer a 'Green Mortgage' to support energy-efficient and solar-fitted homes. Ask your bank to offer such loans; the market is growing.

The CEC's Renewable Energy Program, which began in 1998, makes it even easier, in line with its goal to foster the use of renewable energy and the development of a competitive renewable energy market in California. Residents living in eligible local publicly owned electric utility districts like Los Angeles Department of Water and Power (LADWP) or Sacramento Municipal Utility District (SMUD) can qualify for a Buydown Programme. For PV or wind systems of 10kW hours or less, the Buydown Programme offers homeowners a $4.50 per watt subsidy or up to 50 per cent of the total eligible system cost. Combined with other incentives like state rebates, and federal and state tax credits, homeowners can lower the system cost even further by about 10 to 15 per cent. All this has built an infrastructure which can be tremendously expanded by – and, in turn, give a boost to – the California Solar Initiative (CSI), signed by Governor Arnold Schwarzenegger in 2006. Through the

$3.2 billion CSI programme, offering a variety of incentives, California has set a goal to generate 3000MW of new, solar electricity by 2017.[44]

The Massachusetts Renewable Energy Trust (MRET), established by the state legislature in 1997 and managed by the Massachusetts Technology Collaborative (MTC), is geared up to promote the development and use of renewable energy in the commonwealth. Through programmes like Green Buildings Initiative, Green Schools Initiative, Solar to Market Initiative and Small Renewables Initiative, MTC has provided funds to subsidize the cost of integrating energy efficiency and renewable technologies in homes, businesses and schools in eligible areas. Funding can be used for feasibility studies, design and construction, buydown of system costs and education.

In January 2008, the administration of Governor Deval Patrick and MTC launched 'Commonwealth Solar' to accelerate photovoltaic projects within the commonwealth and spur growth of the Massachusetts solar industry. It provides rebates through a non-competitive application process for the installation of PV projects at residential, commercial, industrial and public facilities. It has $68 million available for funding over four years, starting in 2008, to support an estimated 27MW of PV installations in Massachusetts.[45]

The programmes are gaining popularity and MTC intends to remain innovative in its approach to programme development and funding options so it can reach a wider spectrum of recipients.

These are just two examples of a range of state programmes being launched across the country to promote market development and utilization of renewable energy. Some local utilities, environmental organizations, foundations and banks, too, may offer incentives and special financing options. Contractors and vendors of renewable energy systems usually include this information in their marketing packages and bids. They may also handle the paperwork related to obtaining the upfront subsidies, simplifying the transaction for the customer, which itself is an incentive for customers who may otherwise find the process uninviting.

The sunniest parts of the world are the developing countries and the vast majority of people live in rural areas outside the reach of a grid. These are ideal sites for PV systems. But the potential remains largely untapped. Instead, rural people in developing countries are compelled to buy an expensive, hazardous and polluting fuel like kerosene or diesel, both because that is what is available and because they can buy and pay for it in small amounts on a daily basis. Even though a renewable option may be cost-effective over time, any bigger upfront payment makes it unaffordable for many. So, the real obstacle is not the total cost – they are paying much higher costs for what they use; the obstacle is a lack of innovative financing options spread over time which would make a system affordable now. Such financing options are pivotal. Fortunately, there are some inspiring examples.

One of the pioneering examples of such a financing option was implemented by the rural people of the Puerto Plata province of the Dominican Republic through Enersol, a nonprofit co-op founded in 1984 by Richard Hansen, a former engineer and manager at Westinghouse Electric Corporation, who has played a pioneering role in bringing renewable energy technologies to several Central American countries. The financing option through Enersol permitted matching the monthly payment toward paying off the system over a five-year period with the money saved from the discontinuation of the customer's kerosene expense. People were given the option to pay off their system sooner if they wanted. The financing option has enabled the villagers to install more than 10,000 PV systems since 1985, providing electricity for a variety of needs for homes, businesses, community centres, farms, water purification systems, factories and health clinics. The success has led to more innovative projects and services, and Enersol has been integrated into Soluz, under the umbrella of Global Transition Group, with Richard Hansen at the helm.[46] Enersol's legacy lives as an inspiring story of a personal initiative and grassroots collaboration succeeding in creating some of the world's solar-powered communities!

In Bangladesh, the number of renewable energy programmes launched through NGOs, governmental agencies and commercial enterprises has been growing rapidly in recent years. Among the most notable ones on a massive scale is Grameen Shakti (Rural Power) NGO, the renewable energy division of Grameen Bank, which is the recipient – along with its founder and managing director, Professor Muhammad Yunus – of the 2006 Nobel Peace Prize.[47] Another is the Solar Energy Program for Sustainable Development of BRAC, the largest and the most self-sustainable NGO in the world.[48] The programmes offer options for microcredit financing, in which the two NGOs are world leaders. Rahimafrooz Solar, the solar product marketing division of Rahimafrooz Batteries Ltd, also continues to play a major role in the dissemination of photovoltaic systems through its collaboration with various NGOs and programmes.[49] The country's major battery supplier to solar systems, Rahimafrooz has also launched a praiseworthy battery retrieval and recycle programme for the protection of the environment. Between Grameen Shakti, BRAC, Rahimafrooz Solar and several other NGOs, with partial World Bank support financed through Infrastructure Development Company Limited (IDCOL), over 400,000 small stand-alone PV systems have been installed at rural homes, health clinics, NGO offices, shops, cyclone centres, community centres, restaurants and other places since the late 1990s. Much of this success in their rapid expansion is due to the financing options offered by the programmes.

Some of these programmes are setting global examples. In 2006, Grameen Shakti and Rahimafrooz Batteries were awarded the prestigious Ashden Awards for Sustainable Energy for 'the central roles which they have played in delivering the world's most successful solar power programme bringing light and power to rural people'. Grameen Shakti also received the 'Eurosolar Prize' in 2003 and the 'Right Livelihood Award' (Alternative Nobel Prize) in 2007.

One of the most comprehensive renewable energy programmes in the world, which includes an attractive package of financing and other incentives, has been launched in India. It is a fine example of public-private partnership (PPP). Through the country's Ministry of New and Renewable Energy and its implementing agencies, such as IREDA (Indian Renewable Energy Development Agency), India's renewable energy programme integrates PV, wind turbines, microhydros, solar thermal systems for hot water, biogas plants, and biomass cogenerators and gasifiers. The programme also includes incentives for more efficient traditional wood stoves, the Improved (or *Unnata*) Chulhas.[50]

Microcredit plays a vital role in making renewable energy systems affordable around the world by helping to overcome one of the main barriers, the high upfront cost. At the same time, pointing to the usual range of annual interest rates, 18 to 70 per cent, critics have raised the concern that such interest rates are too high. The lenders usually justify it on the grounds of compensation for default risk and the high cost of administering small loans and providing social-financial support services which often accompany microcredit programmes. The issue, however, is still hotly debated, calling for a fair hearing of both sides and examining those concerns. There is also room for innovation in programme development, efficiency in operation and equitable sharing of benefits. Therefore, to build an infrastructure for expanding renewable energy technologies, as an integral dimension of a sustainable and just economy, policymakers must address these microcredit lending issues by instituting proper controls, support mechanisms and guidelines so that the professed values of microcredit lending are realized. It is encouraging that these concerns are beginning to be raised within the Micro Finance Institution (MFI) industry itself. They should be brought to the forefront for greater public awareness and scrutiny, and acted upon with the utmost urgency.[51]

Incentives like financing options, rebates, subsidies, leasing and tax credits to customers, through public and private sources, are growing around the world. But, on the whole, these still lopsidedly favour nonrenewables – clearly discriminating against renewables. The policies, therefore, must be revised to bring about fairness and the equal treatment of options available to people and businesses. The incentives must be drastically increased for renewables and decreased for nonrenewables. Only this shift will lead to a neutral ground on which any future incentives can be fairly discussed, selected and implemented.

Buy renewable energy credits

Let's assume you do not or cannot generate renewable energy from a system of your own. Can you still help with the transition? Yes, you can! Homeowners, renters, apartment-dwellers, students living in dorms, institutions and corporations are doing so across the country by purchasing renewable energy credits that support energy production from renewable sources.

Here is an example of how the programme works. As a wind farm generates electricity and injects it into the national grid – in addition to reducing the need to generate electricity and reducing pollution compared to the same amount of electricity generated from a fossil fuel-powered plant – it also generates one renewable energy credit (REC) for every 1000kW hours. A range of renewable energy credit suppliers – utilities, nonprofits, businesses – create a contract with the wind farm to sell the credits to interested buyers. The supplier could be your own utility company or a separate one. It adds a little extra to your energy bill, but this is a way to support a cause you believe in. You can access many suppliers through the internet. Purchase what you can afford – every bit counts.

Environmentally conscious businesses are among the major buyers of renewable energy credits. In 2004, Colorado-based White Wave Foods, the maker of Silk soymilk and the country's largest soy foods manufacturer, became the largest American company to purchase wind energy credits for 100 per cent of its manufacturing and operations needs. By 2008 the company's purchase of renewable energy credits reached the equivalent of offsetting more than 204 million kilograms of greenhouse gas emissions annually or taking 40,200 cars off the road for one year.[52] Whole Foods offsets 100 per cent of its energy use in the US and Canada by purchasing over 500 million kilowatt hours of renewable energy credits from wind farms.[53] The extra expense of renewable energy credits is allowing these businesses not only to carry out their social missions but also to attract socially conscious customers. It is good for the business, good for the consumer and good for the planet.

Building green

There's a similarity between the way building design and car design – particularly American car design – evolved over the years. Energy guzzling cars were being designed with little attention to fuel efficiency, if not downright shunning it. Likewise, building designers or architects paid little attention to conservation and efficient use of energy. Schools of architecture and engineering paid scant regard to these topics in their curriculum offerings.

As in almost any situation, there were exceptions. More than 15 years ago I was invited by an organization called Architects for Social Responsibility to speak on the topic of renewable energy and sustainability. The members were a group of architects and engineers who united under a shared vision and with a strongly felt need to address responsibly issues such as sustainability, efficiency, environment, renewable energy and the like within their own profession. In the generally conservative climate of their profession at the time, they were visionaries and pioneers.

The good news is, owing to such initiatives and compelled by the energy and environmental crisis we have encountered, architecture and engineering education and professions are undergoing revolutionary changes. Topics such as 'Green Building' and 'Sustainable Design' are being included in the curriculum and becoming buzzwords in the profession. Out of both choice and necessity, key features of a sustainable design like functionality, aesthetics, economy, resource efficiency and durability are being integrated into a new wave of holistic architectural designs and building practices.

Socially responsible organizations promoting sustainable buildings have emerged and gained mainstream credibility. Among the most notable is the nonprofit trade organization, the US Green Building Council (USGBC). Founded in 1993, with more than 15,000 members, the USGBC has developed Leadership in Energy and Environmental Design (LEED), a rating system to evaluate a building against a series of standards and assign points for qualifying it as a 'green' building. Energy and water efficiency, renewable energy, land use, waste handling, indoor air quality and environmentally preferable materials are some of the standards.

LEED certification is not yet a universal requirement for building contracts, but with growing environmental concerns the certification is gaining that credibility.[54] It is applicable to private homes as well as institutional buildings. The highest rating a 'green' building can get is Platinum. Bruce Oreck's 'Next West' home in Boulder, Colorado, is a pioneering LEED Platinum, net-zero-carbon house in the

US.[55] Some exciting examples of larger green buildings are the Michael E. Capuano Early Childhood Center in Somerville, Massachusetts,[56] the Ordway building of the Woods Hole Research Center in Woods Hole, Massachusetts,[57] and the Leslie Shaoming Sun Field Station at Stanford University's Jasper Ridge Biological Preserve.[58] These buildings are projecting a new horizon for the professional future of architects, engineers and builders. Mandating the use of renewables in building codes would not only facilitate the transition, it would also add a whole new revolutionary and sustainability dimension to their education and professions.

Public interest in green buildings is growing rapidly. Building Energy, the annual conference and trade show of renewable energy and green building professionals, held in Boston and organized by the New England Sustainable Energy Association (NESEA), is drawing record crowds. It has quickly evolved into one of the nation's major and most successful conferences on the subject, attracting both professionals and interested visitors from around the world.[59] Whoever said 'People build buildings, then buildings build people' had it right. So, when we build a building – let's build it green!

Solarizing academia

Educational institutions must become more proactive in the transition. Unquestionably, in schools, colleges and universities around the world, programmes, projects, courses, research and other activities in renewable energy have been growing. The subject is still on the fringe, but the trend is certainly growing.

Some institutions have taken the subject to a practical level. Georgetown University in Washington, DC, was one of the first educational institutions in the country to set such an example. In 1982 a PV system of over 3000m² was installed on the roof of the university's Intercultural Center. In addition to being a great educational resource for both the university's students and visitors, the system generates about 40 per cent of the building's electricity needs and saves $45,000 annually in energy bills for the university.[60] At the University of South Florida a charging station with a roof of PV modules charges a fleet of the university's vans and also serves as a parking shed.

With the vision of a carbon-neutral campus, the Lewis Environmental Studies Center of Oberlin College, Ohio, sets an inspiring example. The academic facility and the adjacent parking pavilion are covered with building-integrated 59kW, 325m² and 100kW, 818m² PV arrays, respectively. Constructed during 1998–2000, the PV system generates enough electricity to turn the centre into a net energy exporting facility.[61]

In the Greater Boston area, the hub of educational institutions, several universities – Harvard, MIT, Tufts, Boston and Northeastern – have installed grid-connected PV systems on their buildings. There are other universities, colleges and schools across the US that have installed PV or wind turbines or both on their campuses.

WGBH, a PBS-affiliated broadcasting facility located in Brighton, Massachusetts – not an academic institution, but certainly an educational institution par excellence – in 2006 became the first network TV station in the country to go solar. The LEED-certified building with several green features has a 100kW rooftop PV system, surrounded by a roof garden. A custom data-acquisition system in the building's lobby explains and monitors the system's performance, while also serving as an educational tool for visitors and demonstrating WGBH's commitment to sustainability.[62]

Solarizing academia can begin early. Solar Now, a project based at Beverly High School, Beverly, Massachusetts, installed a 100kW PV system during the early 1980s and a 10kW wind-turbine during the 1990s. The project has not only been a research and educational resource for the high school for years, but also for a growing number of interns and visitors from around the world.[63]

On 3 June 2004, at the John D. O'Bryant School of Mathematics and Science, Mayor Thomas Menino inaugurated a 2kW PV system installed on the southeastern wall of the high school. The project resulted from a collaborative effort among its partners, including the Boston Public Schools, the City of Boston, the MIT Space Systems Laboratory, the MIT Edgerton Center, Massachusetts Technology Collaborative, Heliotronics, Massachusetts Energy Consumers Alliance and Solar Boston. It has offered a great opportunity for the students to learn about concepts in renewable energy, physics, engineering and environmental science through hands-on experience with the PV system.[64] What is most inspiring is that in 2009 a group of students, under the guidance of Steve Fernandez, their physics teacher who also spearheaded the school's solar project, is taking what they have learned about the technology to help install a solar project at a health clinic in a rural Mayan village in Guatemala. Indeed, through their education in renewable energy, the students are living up to the name of the organization they have formed, Students Without Borders! Helping with the educational aspect of it has been for me a deeply gratifying and inspiring experience.

At Dhaka University in Bangladesh, the Energy Park, with some stand-alone PV systems and solar box cookers, was established during the early 1990s for demonstration and research purposes, and has become a place of growing public attraction. The Energy Park also holds the Energy Fair, an annual event bringing together students, educators, the general public, public officials, and vendors and users of a wide range of renewable energy technologies from all over the country. It is a project of the Renewable Energy Research Centre (RERC) under the university's Department of Applied Physics, Electronics and Communication Engineering. In 2007, RERC also added a successfully implemented 1.1kW rooftop grid-connected PV system to its repertoire of projects.[65]

Projects like these not only generate clean energy and serve as learning labs and demonstration models, they also constitute an action that practices what the schools, universities and other educational institutions preach, or teach!

Stewardship

Growing environmental concerns due to the use of nonrenewables has been having a profound impact on many churches and other religious institutions. Beyond prayer and worship, members are coming together to take the matter into their own hands with an active response which addresses the issue of energy directly.

In March 2000 the Eco-Justice Working Group of the National Council of the Churches of Christ in the US published 'Energy Stewardship Guide for Congregations'. It reminds us of the biblical calling for caring and protecting all of creation. Citing society's use of energy, which is causing many problems for human beings and the non-human aspects of creation, it outlines some specific steps congregations and other faith-based institutions can take to use less energy and turn to renewable energy sources.[66]

The 1997 General Convention of the Episcopal Church USA passed a resolution calling on members to practise energy efficiency. The Rev. Sally Bingham of Grace Cathedral in San Francisco and Steve MacAusland of St Paul's Church in Dedham, Massachusetts, paved the way for the founding of Episcopal Power and Light, a national ministry devoted to environmentally responsible energy usage throughout the Episcopal Church. It is now renamed Massachusetts Interfaith Power and Light (MIPL) to reflect an expanding stewardship mission and broader discussion of energy scarcity, security and justice. 'After love and the Holy Spirit, it is energy that makes the world go "round",' says MacAusland.[67] So, as Tom Nutt-Powell, a member of All Saints Parish in Brookline, Massachusetts, and a founding director of MIPL, puts it: 'Our intent is to do everything in our power to help our congregations and their members learn to practice energy conservation, invest in energy efficiency, pool our purchasing power in the marketplace and so buy our energy – electricity, oil, and gas – in bulk so that we can save enough money to afford those slightly more expensive renewable energy products.'[68]

To spread the message and honour Earth Day, in April 2002 MIPL convened a conference that brought together people of various faiths to address, with an appeal for action, the responsibility which we as humans bear as stewards of the Earth. The conference inspired several action-oriented projects including an energy audit for buildings, better insulation, and use of energy efficient fixtures.

The aforementioned are examples of the many calls to stewardship sprouting not just in the US but around the world.[69] They are prompted by the critical global situation we find ourselves in and the alarming future we are marching towards. Yet, they are reminders of a sense of stewardship which, for a long time, has been a part of the human spiritual tradition across all cultures. It has been said in the following proverb, the source of which has been attributed to Kenyan, Native American and Shaker traditions: 'Treat the Earth well. It was not given to you by your parents, it was loaned to you by your children.'

Policy-programme-practice (PPP)

No issue is of greater public concern and responsibility than the energy issue. Rightfully, a public policy on energy, which both expresses and influences the will of the people, is of critical importance. And it's good news that more and more countries around the world are recognizing the importance of renewable energy and making it a part of their national policies.

Adding renewable energy to the energy policy mix is good news, and the more the better. However, to make a transition to the renewable energy path, which implies a fundamental reversal of our entrenchment in the nonrenewable energy path, requires more. It can no longer just rely on chance or manipulated market forces which are geared to profit fossil fuel and nuclear industries. A pro-renewables, dynamic, comprehensive and integrated policy – with a policy-programme-practice continuum – must be conceived and implemented. Applicable to every country in the world, from community through national levels, it will require the following:

1 Maximizing conservation and efficiency in the use of nonrenewables, while utilizing them only as transitional resources;
2 Transparent, equitable, and socially and environmentally responsible public-private partnership to maximize efficiency and expediency;
3 A moratorium on further entrenchment into the nonrenewable path, combined with disincentives such as a reduction of subsidies for nonrenewables and a progressive carbon tax and cap on carbon production;
4 Proactive and massive utilization, investment and development of appropriate renewable options;
5 A combined offering of public education, technical support, renewable energy programmes, a legal framework and financial incentives to renewable energy users and producers;
6 Collaboration between experts and stakeholders in both renewable and nonrenewable energy fields to devise an integrated and comprehensive public policy – holistically assessing both the renewable and nonrenewable options, from both global and local perspectives – to lead the transition through action (programmes into practice).

It may seem like a daunting task, but insisting on a path which has led the world to the crisis we are in is not a solution; it is suicidal. On the other hand, the revolutionary scope of renewable energy offers us an alternative option, a hope, an opportunity – in which anyone across the world, in small and big ways, can participate and contribute – to accelerate the transition. If we choose to act, we have the technology, experience and inspiring examples.

In this, Germany stands out. Its policy is comprehensive and decisive. In 2000, to phase out its 19 nuclear power plants over the next 20 years. Thereby, it joined the ranks of Austria, Belgium, Italy, the Netherlands, the Philippines and Sweden, who have the policy to phase out nuclear power as a dead-end technology.[70] But let's not be naive. As the energy crisis deepens, along with a growing concern over climate change, there will be pressures and manipulations of public sentiments by the nuclear industry to reverse some of these decisions. Yet that will happen not because nuclear power has proven to be any more useful or less destructive. It will happen because the critics of nonrenewables have weakened their cause by not being persuasive enough about the potentials of renewable alternatives, not just in words but through actions. Germany's renewable energy policy is progressive, visionary and based on actions. Let's hope the policy sustains, expands and continues to inspire the world.

Thanks to the alliance of the Social Democratic Party and the Green Party, and the visionary leadership of parliamentarian Hermann Scheer, the German government is gearing up for a transition to a 100 per cent renewable energy economy. It is the world's most aggressive policy for transition into the renewable energy path, marking the period until about 2020 as the 'make-or-break' years for the transition.[71] 'Nuclear power and fossil fuels are the choices of the past. Renewable energy is the choice of the future that is here today,' says Hermann Scheer, Chairman of EUROSOLAR, General Chairman of the World Council of Renewable Energy (WCRE), President of the International Parliamentary Forum on Renewable Energies, member of the German Bundestag, and author of *A Solar Manifesto* and *The Solar Economy*.[72]

Policy supports action, while action speaks for policy. Barely starting in 1990, fuelled by a synergistic interplay of policy and action, since then the installed wind capacity has grown by more than 2000 per cent, biomass by more than 500 per cent and photovoltaic installations by more than 15,000 per cent in Germany.[73] Germany today is one of the leading producers of both wind energy and photovoltaic electricity in the world.

To accelerate the transition, projects of a wide range of designs and scales are being encouraged with highly attractive incentives. Homeowners, communities, farms or others choosing to install a grid-connected system can sell the electricity for as much as 4.5 more than what they would pay for buying electricity the conventional way. The incentive is termed Feed-in Tariff (FIT), and it has been made into a law. Many homeowners, therefore, have turned their roofs into an income-generating rooftop solar energy industry! Germany passed a Renewable Energy Act with these key elements: legally guaranteed access to the grid for each power supplier of any size from renewables, a guaranteed profitable income from investments in renewable systems, and no limit to the quantity of energy a company can produce from renewables so the company can base its investments on long-term prospects, such as more profitable returns from mass production. The Renewable Energy Act has essentially launched an autonomous renewable energy movement not subject to the control of, or obligations to, the conventional power companies. The public policy and private entrepreneurs are partners with each other. Its success is inspiring: the goal of 3000MW annually of new renewable installations without large hydropower has been exceeded by 20,000MW in six years, while also creating 150,000 new jobs in the renewable energy sector.

Cuba is another example. The collapse of the Soviet Union in the 1990s led to Cuba's loss of access to Soviet oil, chemical fertilizers and pesticides, creating an unprecedented energy, food and agricultural crisis. But the crisis also opened a window to an opportunity: to quickly transition out of oil- and chemical-based agriculture and get into organic farming throughout the country. Cuba seized the opportunity and within the next few years practically all the arable lands in Cuba were transformed into organic farms and urban gardens. Today, 50 per cent of the vegetables eaten in the city of Havana are organically grown within the city limits and nationally 80 per cent of the food is grown on organic farms. In addition, through CUBASOLAR, the Cuban society for the promotion of renewable forms of energy and respect for the environment, Cuba has installed photovoltaic systems in schools across the country. Combining policy and practice through joint efforts of private ownership, cooperatives and state programmes, it is an inspiring story of Cuba's survival of 'peak oil' and a hopeful example for the rest of the world.[74]

Pope Benedict has gained the reputation as the 'Green Pope' for his outspoken voice of stewardship against environmental destruction, labelling pollution a sin. Around the world he is conveying his message of stewardship. On 17 July 2008, speaking at the 23rd World Youth Day in Sydney, Australia, before a crowd of over 150,000, he said: 'My dear friends, God's creation is one and it is good. The concerns for non-violence, sustainable development, justice and peace, and care for our environment are of vital importance for humanity.'[75] Acting on these concerns, in 2008 the Vatican, a 110-acre city-state with a population of about 900 but visited by millions, has installed a solar system on the roof of its main auditorium, the Nervi Hall. The 2400-module system covering 5000m^2 roof space will generate 300MW hours of electricity annually. It will offset about 225 tons of CO_2 emissions and save the equivalent of 80 tons of oil each year. The $1.6 million system was donated by German companies SolarWorld and SMA Solar Technology.

The Vatican is already planning to build a much larger solar power plant on a 3km^2 site it owns north of Rome where the transmission centre for Vatican Radio is currently located. The proposed solar plant will generate six times the energy needed to power the transmission centre, so the surplus energy will be fed into the Italian national grid for distribution among its customers.[76]

Islands are the places most vulnerable to rising sea levels. As mentioned in Chapter 4, the Maldives, a republic with a cluster of 200 small inhabited islands in the Indian Ocean, faces a precarious future, but rather than merely succumbing as a victim, it is taking ownership of the solution. This also puts the nation in a much stronger moral position to demand that the rest of the world should act on solutions, too, without which its own efforts will be futile. In March 2009, the Maldives' President Mohamed Nasheed announced a plan to become the world's first carbon-neutral country by generating all its electricity from the nation's plentiful renewable resources – sun and wind. The $1.1 billion plan involves installing 155 1.5MW wind turbines and a 0.5km^2 solar farm to meet the country's energy needs. It's a sizable cost for the nation, but any external funding support which can and should be extended to it will be an investment not only in the Maldives' own survival from an impending threat, but also in the eventual survival of the world.

With its lush tropical landscapes and pristine beaches, the Maldives is also a growing tourist attraction. But this economic advantage comes with the price of accelerated greenhouse gas emissions from air travel and from commercial activity on the island. To offset the impact the government has planned to buy EU carbon credits.[77]

Samso, a Danish island located 25km from the mainland, was entirely dependent on imported oil and coal only a decade ago. It is now considered the most energy independent place in the world. Supported by Denmark's progressive pro-renewable energy policy, the island with a population of 4300 generates more than 100 per cent of the electricity it needs from wind power and 70 per cent of its heat for hot water and district heating from solar thermal systems and biomass. Conservation is not a compromise, it is prized. As Soren Hermansen, a former farmer and environmental studies teacher, and also the main force behind the island's transition, put it: 'To us, going for lower energy is like a sport.'[78]

Japan is the number one producer of electricity from photovoltaics in the world. But it is not stopping there. It is continuing to devise and implement a variety of programmes to accelerate its use across the country. Ota, Japan's 'Solar City', is an example. Located 80km northwest of Tokyo, in the Pal Town neighbourhood, three-quarters of its homes have PV systems. What is special about this programme is that all the 550 4kW grid-connected systems were given free to the residents by the government to ensure a steady energy supply and avoid blackouts.[79]

All eyes are turning to the birth of a fascinating experimental city in the middle of the Arabian Desert. It is named 'Masdar', meaning 'the source' in Arabic. The seven-year $20 billion plan, when completed, will make Masdar – spread over 6km² and powered mainly by renewables – 'the first city where carbon emissions are zero'. Its 50,000 people will meet most of their energy needs from renewable sources. Photovoltaics, wind turbines, solar thermal systems, biomass and solar hydrogen will be combined to power what the planners claim to be the 'city of the future'. The city will be free of private cars. Public transportation will be provided by a rapid transportation system of solar-powered car-like vehicles. In the midst of a desert, Masdar will have a blooming landscape of greeneries. A wide range of conservation and efficiency measures will be instituted.[80]

Is Masdar the 'city of the future'? Will it deliver on all its promises? There are both believers and sceptics. Perhaps the answer lies somewhere in-between. What is also highly significant is that Masdar is being constructed in Abu Dhabi, the oil-rich capital of the United Arab Emirates. There is an acknowledgement – and a message – here: the age of oil is coming to an end. To assure a sustainable and prosperous future, we must turn to the Sun.

Iceland is rapidly progressing towards its goal of becoming a 100 per cent renewable-energy-powered nation by 2050. An island of 103,000km² located in the mid Atlantic Ocean, with a population of around 320,000, its geology is made up of hundreds of volcanoes, geysers and hot springs – prime sources of geothermal energy. Heated rocks and ground water are so near the surface – some even above the surface – that geothermal systems can be installed with minimal drilling and disruption of the natural geological formation. The island also has many glaciers and rivers – prime hydropower sources, which can be trapped without disrupting the ecology. Together, geothermal and hydropower sources provide 70 per cent of Iceland's total energy need and 100 per cent of its electricity and heat.

Iceland's fishing fleets and transportation sector still rely on imported oil and coal, but it has already set a policy to power these vehicles by hydrogen fuel cells produced by geothermal electricity. Intensive research and successful demonstrations in cars, buses and ships are paving the way. By 2050 Iceland is poised to become the world's leading renewable-powered hydrogen economy.[81]

Like the rest of the world, Iceland too is hit by the global economic recession – and, especially due to its huge investments in external European financial markets, much more so than many other countries. However, the country's reliance on domestic and renewable energy sources is also its best hope, as well as an economic foundation, for its recovery.

Now to the United States. I want to elaborate on this section simply because the policies of the US have the most impact, not only on itself, but also around the world. In addition to the US being the largest consumer of energy in the world, its policies are also crucial internationally because they have direct and indirect impacts on the world economy and on two of the major financial institutions, the World Bank and the International Monetary Fund (IMF). These institutions, in turn, dictate most of the 'development' patterns and programmes throughout the developing world. Additionally, the flip-flop commitment of the US government to renewable energy over the past few years simply cannot be repeated if we are serious about the renewables transition. One way to guard against that kind of flip-flopping is to have a historical understanding of this process.

With 5 per cent of the world's population, the US consumes 24 per cent of the energy produced in the world, twice the per capita consumption of energy used by other industrial countries with comparable material standards of living. Consequently, the US is also responsible for 25 per cent of CO_2 emissions affecting the global climate. And to meet its energy needs, both in policy and practice, the US has favoured the costly path of external dependence. The US imports more than 50 per cent of the oil it consumes. This has led to the need to secure an oil supply from the Middle East – spending billions of dollars annually – with guns, missiles and human lives rather than tapping into the free and inexhaustible renewable energy sources at home.

Some hopeful changes began during the Bill Clinton administration. These changes were especially significant because they were taking place following what can be called an 'anti-renewable energy era' in the nation's history. Funding for renewable energy research, development and demonstration projects declined by 80 per cent between 1980 and 1993. The renewable energy industry was crippled. An estimated 80 per cent of the renewable energy technology manufacturing companies in the US either folded or went overseas, resulting in the country's fall from its position as the leading manufacturer of photovoltaics in the world to trailing way behind Japanese and European manufacturers who have risen to global dominance since. Symbolizing the 'anti-renewables' era, the solar panels for hot water put up on the White House roof during the Carter administration were actually taken down during the succeeding Ronald Reagan administration and sent off to inactive exile in the General Services Administration (GSA) warehouse in Franconia, Virginia.

Some good news is that renewables have returned to the White House! Initiated by the National Park Service – not the White House administration – one PV and two solar thermal systems were installed during the summer of 2002. Solar Design Associates, headed by its president and a solar pioneer, Steven Strong, designed and oversaw the installation by Aurora Energy of a 10kW Evergreen Solar photovoltaic system. The utility-connected system is placed on the roof of the main building used for White House grounds maintenance. One of the solar thermal systems heats the pool and spa and the other provides domestic hot water. James Doherty, the architect and project manager at the National Park Service Office for White House Liaison, said: 'We believe in these technologies, and they've been working for us very successfully. The National Park Service also has a mission to lower our energy consumption at all our sites, and we saw an opportunity to do this at the White House grounds.'[82]

More good news is that the solar panels removed from the White House were liberated in 1992 from their exile and for 12 years provided hot water for the cafeteria of Unity College in Unity, Maine. Some of them were also used for student experimentation and education. They are now preserved as historical artefacts. The panels were acquired by the small college, located in central Maine and dedicated to environmental education, under the government surplus donations programme.

In 1991, Peter Marbach, Unity's director of development, filed an application with the GSA to acquire the unused solar panels. Following payment of a $500 fee by the college, the GSA donated the panels. As the story goes, Marbach then removed most of the seats from an old school bus belonging to the college and drove down to Virginia to pick up the panels.[83]

Under the Clinton administration, the Department of Energy (DOE) budget for energy efficiency and renewables was increased for fiscal year 1994. Though miniscule compared to the $30 billion a year subsidy to the fossil fuel and nuclear industries, it was a critical turning point, nevertheless. The trend continued through the administration's two terms in office through 2000, with the allocation for efficiency and renewables reaching about $1 billion.

The most significant federal commitment to renewable energy since the tax credits and increased research and development funding of the 1970s was the Million Solar Roofs Initiative (MSRI), announced by President Clinton in June of 1997 in a speech to the United Nations Special Session on Environment and Development in New York. Crafted by the DOE in cooperation with the Solar Energy Industries Association (SEIA), the MSRI called for the production and installation of 1 million residential and commercial solar systems in the US by the year 2010. The technologies selected were photovoltaics for electricity and solar thermal systems for hot water. The $200 million/year programme included low-interest purchase options for customers, the creation of 100,000 jobs, a Solar Home Registry, federal procurement of solar systems for federal buildings and federal support for the US solar energy industry. It also supported the forming of regional partners around the country to promote renewables.

The initiative stirred up an interest in renewables both in the public and private sectors. Individual states and cities found an ally and some support for their own efforts and initiatives, as did the investors and industries in the private sector. At the same time, it also brought to the surface the need for the Department of Energy officials to be able to distribute the money more efficiently, to build bridges between policymaking and an understanding of how the real world works so that the policies can be relevant and implemented effectively, and to rethink an overemphasis on marketing renewable energy technologies without enough emphasis on subsidizing demonstration projects. Such demonstration projects serve to both educate and inspire people on a massive scale to invest their own resources in renewable technologies about which they had very little knowledge and even were sceptical about. In 2007 MSRI changed its name to Solar America Initiative (SAI), with some modifications in its mission and programmes. Hopefully the lessons from MSRI will be carried forward, too.

Another significant federal commitment is the Senate's passing of the Renewable Portfolio Standard (RPS) in July 2003. The RPS requires major electric utilities to obtain a minimum percentage of their electricity from new renewable sources by a certain date. Some states have been at it individually for some time and by 2009 about 30 states, including the District of Columbia, had adopted the RPS. Some other states have set their own renewable energy goals or are considering the RPS.

These are no giant steps but, at least, these are small steps by a giant towards the transition.

But even that little progress was threatened by the National Energy Policy of the Bush-Cheney administration. The highlights of the policy were: drilling for oil and gas in the Arctic National Wildlife Refuge in Alaska, offshore drilling for oil and gas, expedient expansion in the number of fossil fuel and nuclear power plants, weaker safety standards to streamline nuclear power plant licensing and renewal of the law that shields nuclear power plants from liability, and even withdrawal of regulations that prevented the restarting of power plants that had been shut down for environmental concerns. The administration was simply playing out its subservient role under President George W. Bush and Vice President Dick Cheney, both with a deeply vested interest in oil. Cheney famously declared that 'conservation is a personal virtue', therefore it cannot be a basis for a national policy. They proposed major cuts in energy conservation, efficiency and renewables, while billions of dollars were funnelled into subsidies for nonrenewables. Also, the administration continued to refuse to sign the Kyoto Protocol, the UN treaty signed in Kyoto in 1997, which asks nations to reduce greenhouse gas emissions to around 5 per cent below 1990 levels by 2010. At the World Summit on Sustainable Development in 2002 the proposal to establish a global goal to achieve 15 per cent of worldwide energy production from renewables by 2015 – strongly supported by Europeans and developing countries – was strongly opposed by the US delegation. Last, but not least, is the aggressive launching of the Iraq War for oil. Based on the lie of disarming the country of weapons of mass destruction (WMD), and causing the death of nearly 5000 US soldiers and an estimated 1 million Iraqis (and counting), the war itself has proven to be the deadliest WMD of our time.

A detailed record of the Bush administration's environmental legacy is in the book, *Strategic Ignorance: Why the Bush Administration is Recklessly Destroying a Century of Environmental Progress*, co-authored by Carl Pope, Executive Director of the Sierra Club, and Paul Rauber, a senior editor of *Sierra* magazine.[84] As to the administration's commitment to renewable energy, a *New York Times* editorial put it well: 'In an age when people are worried about global warming and the country's growing dependence on imported oil, the federal effort on alternative energy sources is pathetically small.'[85]

Fortunately, the National Energy Policy continued to draw fire. It was being criticized not just by environmental groups and other supporters of renewable energy, but also by the general public – across party lines and around the world – who were waking up to its potential and universal destructiveness. People were beginning to see the connections between nonrenewable energy, environmental destruction, economic crisis and war. Disillusionment with the administration was setting in, evidenced in the Democratic turnover in the House of Representatives and the Senate. It was showing in the polls, protests and demonstrations. In the growing worldwide peace marches opposing the Iraq war some posters and placards demanded 'No Blood for Oil', heralding a growing public interest in renewables. I marched with one I made, 'Go Solar for Peace!' The Bush-Cheney government increasingly became isolated and even an embarrassment to its own Republican Party. Behind the façade of distancing itself from its own party predecessors, but basically trying to push the same agenda (John McCain's call for 'Drill, baby, drill!', etc.), the Republican Party made a futile effort to become relevant and win the presidential election with the McCain-Palin ticket. The rest is history.

With President Barack Obama and his administration there's a renewed hope for change. In his inaugural address the president promised to 'harness the sun and the wind and the soil' and announced in later speeches, 'To spark the creation of a clean energy economy, we will double the production of alternative energy in the next three years.' Speaking at Georgetown University, Washington, DC, on 14 April 2009, he remarked: 'The third pillar of this new foundation is to harness the renewable energy that can create millions of new jobs and new industries. We all know that the country that harnesses this energy source will lead the 21st century.' Since then he has also called for conservation, efficiency, anti-pollution measures and capping carbon emissions. These words are influencing some of the administration's agenda.

In April 2009, Energy Secretary Steven Chu announced the creation of a team to distribute over \$40 billion for renewables contained in the president's economic stimulus package for energy projects. At a January 2010 Senate hearing, Interior Secretary Ken Salazar, advocating for a massive deployment of solar energy on public lands, said: 'For the first time ever, environmentally responsible renewable energy development is a priority at this Department.' All these announcements and more come on the backdrop of Al Gore leading a bold proposal to combat climate change and free the US from its dependency on foreign oil by making a major transition to renewable energy sources within ten years. 'Today I challenge our nation to commit to producing 100 per cent of our electricity from renewable energy and truly carbon-free sources within 10 years,' said the former vice president before an audience at the Daughters of American Revolution Constitution Hall in Washington, DC, on 17 July 2008, and reiterating it at his Meet the Press interview the following Sunday.[86]

To succeed, these promises and proposals will require fundamental policy changes geared for action along with massive public education and participation. Conventional 'top-down' and 'bottom-up' approaches must unite in collaborative actions. Without a doubt, there will be a growing opposition. Those with vested interests in nonrenewables are still very powerful – the economy and politics still remain hostages to their pervasive control – and an energy crisis is a record boon to their profitability. Proposals and policies will be blocked and promises will tend to be compromised beyond their effectiveness for a transition. Transitional policies and steps will be blamed for continuing economic recession and growing unemployment. There will be attempts to actually undo such policies and steps – we have seen that happen during the previous administration. This time recession and unemployment will be used as special weapons and excuses. There will be social, cultural, political and economic issues and obstacles at grassroots levels, which will require a skilful and persuasive response. We need to understand that the current economic recession and unemployment are consequences mainly of past policies – both Republican and Democratic – and the long legacy of our aggressive entrenchment into the nonrenewable energy path. Even during sincere attempts to solve these problems, the consequences will continue to be felt for some time.

In the face of these challenges, the true integrity and strength of the political promises will be tested. The beginnings of the Obama administration – which has seen the president's declaration during his first State of the Union address on 27 January 2010 that, along with reiteration of his renewable promises, we also need to build additional nuclear power plants and allow offshore drilling, as well as an escalation of the war in Afghanistan – have shown that the test is on. There is cautious optimism, at best, as to how much of the president's promises, and the Democratic Party's campaign pledges of economic reforms to environmental restoration to global reconciliation through constructive dialogue, will be translated into successful policies and practices, substantial enough to expedite the transition toward a sustainable future. It behooves us to remind ourselves that the solutions lie not in re-entrenching ourselves in the same nonrenewable energy path of depleting nonrenewable resources, environmental destruction, economic collapse, militarization and war, but in getting off that dead-end, suicidal path as quickly as possible. If we fail to do this, then Gulf spills, Chernobyls, natural gas well explosions and coal mine collapses can no longer be rationalized as 'accidents,' but predictable consequences of human ecological stupidity, political expediency, technological limitations, hubris and greed. It also reasserts the urgent need for massive public education for action, not just reliance on solutions promised or to be sanctioned top-down by any president or political administration. Ultimately, it's public demand, with action, which moves the politicians and their politics and policies – that's the power of the people! Fortunately, there are promising signs – locally, nationally, internationally – of growing public movements demanding to solve the climate crisis and choose renewables.

The movements must be accelerated. The march for peace must continue. Out of that true and democratic political process will evolve a sustained public policy, with profound implications for both the US and the world.

6

An Act of Dialogue

The highest wisdom has but one science, the science of the whole.

Leo Tolstoy[1]

Perhaps at no other time has wholeness been so difficult, so nearly out of sight, and therefore the need to address ourselves to it is so crucial.

George W. Morgan[2]

We are born whole. As we grow, all too often, somewhere along the way we get divided, not as branches with fruits and flowers of a common tree that sing out our natural diversity, rooted in unity and wholeness, but as cut or fallen branches that in their artificially disconnected existence fail to recognize themselves or each other. Not only is our body compartmentalized, our psyche is too. Unless we somehow succeed in retaining or regaining that consciousness as whole human beings, the dissolution impacts the way we treat other fellow human beings, the way we treat other nations, the way we treat other species and the way we treat our common home, the Earth. Alienated from each other, and from ourselves, even our natural instinct for survival gets driven by fear, hostility, insecurity, exploitation, and either incapability or unwillingness to face the other – to engage in dialogue. These tendencies do not just reside in our psyche, they are projected in the way we conduct practical affairs. The divisions get institutionalized, even defended.

The energy crisis – compounded by all its catastrophic consequences – ruthless exploitation of the Earth's natural resources, climate change, poisoned air, water and soil, collapsing economies and wars – is a symptom of the fundamental crisis of our wholeness. To solve the energy crisis in a sustainable way, we need to face the challenge of regaining our wholeness and reconnect to Nature – as an integral element of Nature – and with this holistic perspective, engage in the fundamental act of dialogue with each other, committed and fully aware that there will be barriers along the way.

A barrier is the classic schism that exists between conferences, academics, studies, policies, plans and recommendations on the one hand, and action, practice, implementation, accountability and results on the other. All these elements are needed and there is no reason to undermine the importance of any dialogue. But elements have to unite and thought and action have to combine to be effective. All too often, driven by narrow specialization, compartmentalized perceptions and petty competition, within as well as between parties, the schism is reinforced. Eventually, it is wasteful, illusory and self-defeating for all. Generally, conferences breed more conferences rather than a solar roof, just as intellectually deficient action produces more frustration and resignation than results.

Transitioning to the renewable energy path is necessary for our survival – physically, economically, environmentally, politically and, no less, morally. It is a choice and a task before all of us. There are many aspects and pieces to the task and none of us can claim the mastery of or need to do it all. It is similar to the famous Indian fable about the seven blind men who went to see the elephant. The one who touched the body thought the elephant looked like a wall, another touched the ear and thought the elephant looked like a fan, another touched the tail and thought it looked like a rope, and so on. So, they all were correct in their perception of the part of the elephant they touched, but what the elephant really looked like rested with seeing the elephant whole. The moral of the story, this holistic, gestalt view of the necessity of seeing the whole, and the whole being more than the sum of its parts, is illuminating and educative.

Whatever our orientation is – economic, environmental, social, scientific, medical, educational, industrial, technical, aesthetic, literary, legal, political, historical, philosophical, moral, spiritual – we all see a part that's real, we all have something to offer and we all have something to gain. The multifaceted energy crisis will have to be understood and addressed from multiple perspectives and skills. Bridges have to be built between policymakers and practitioners, economists and environmentalists, academics and field workers, specialists and generalists, educators and activists, idealists and realists, industrialists and ecologists, programme planners and implementers, problem researchers and problem solvers. Some bridges exist, but more as exceptions than the rule. Bridges have to be built within a country and with the world. It's good for a country, and it's good for the world. Wherever we live, we breathe the same air, we drink the same water, we stand on the same soil and we are sustained by the same Sun. The climate crisis both affirms and warns us that we cannot escape our interdependence. The only sustainable solution is a global solution. This is not just philosophical speculation; it is the core of the science of ecology. We must be willing to embrace this interconnectedness with mutual appreciation and responsibility, to celebrate this unity, and to engage in a dialogue to bring together our hopes, perceptions, values, knowledge, initiatives and expertise into a synergistic and holistic vision – and turn this vision into both individual and collaborative actions, with the utmost urgency. The potential of the enormous range and variety of renewable energy options already available to us gives me the reason and hope that such actions are possible. And it's a choice – at once personal, as well as global. Therein lies the possibility of a successful, revolutionary transition – a transition to a world of sustainability and enduring peace.

Notes

Preface Racing for Survival:
Transitioning to a Renewable Energy Path

1. Edison, Thomas Alva (1931) quoted in *Power for the 21st Century*, Solar Electric Light Company, Chevy Chase, MD. Also, http://radio.weblogs.com/0109157/stories/2003/01/20/randomQuotes.html, accessed 5 February 2010
2. President Barack Obama, from the Inaugural Address, 20 January 2009
3. Freese, Barbara (2003) *Coal: A Human History*, Perseus, Cambridge, MA
4. Editors of *The Ecologist*, *Blueprint for Survival* (1972), New American Library, New York, NY; Berger, John (1977) *Nuclear Power: The Unviable Option*, Dell, New York; Byron, Michael P. (2006) *Infinity's Rainbow: The Politics of Energy, Climate and Globalization*, Algora Publishing, New York, NY; Commoner, Barry (1976) *The Poverty of Power*, Alfred Knopf, New York, NY; Campbell, Colin (2004) *The Coming Oil Crisis*, Multi-Science Publishing Company and Petroconsultants SA, Essex, UK; Campbell, Colin J. and Laherrère, Jean H. (1998) 'The End of Cheap Oil', *Scientific American*, March, pp78–79; Deffeyes, Kenneth (2001) *Hubbert's Peak: The Impending World Oil Shortage*, Princeton University Press, Princeton, NJ; Deffeyes, Kenneth (2005) *Beyond Oil: The View from Hubbert's Peak*, Hill and Wang, New York, NY; Flavin, Christopher and Lenssen, Nicholas (1994) *Power Surge*, Norton, New York, NY; Goodell, Jeff (2007) *Big Coal: The Dirty Secret Behind America's Energy Future*, Houghton Mifflin, Boston, MA; Goodstein, David (2004) *Out of Gas: The End of the Age of Oil*, Norton, New York, NY; Hartman, Thom (1999) *The Last Hour of Ancient Sunlight*, Harmony Books, New York, NY; Heinberg, Richard (2003) *The Party's Over: Oil, War and the Fate of Industrial Societies*, New Society Publishers, Gabriola Island, BC, Canada; Kunstler, James Howard (2006) *The Long Emergency: Surviving the End of Oil, Climate Change, and Other Converging Catastrophes of the Twenty-First Century*, Grove Press, New York, NY; McKillop, Andrew, with Newman, Sheila (eds) (2005) *The Final Energy Crisis*, Pluto Press, London; Meadows, Donella H., Meadows, Dennis L., Randers, Jorgen and Behrens III, William W. (1972) *The Limits to Growth*, Universe Books, New York, NY; Merz Sr, Kenneth (2008) *Living Within Limits: A Scientific Search for Truth*, Algora Publishing, New York, NY; Mesarovic, Mihajlo and Pestel, Eduard (1974) *Mankind at the Turning Point*, Dutton and Reader's Digest Press, New York,

NY; Roberts, Paul (2004) *The End of Oil: On the Edge of a Perilous New World*, Houghton Mifflin, Boston, MA; Scheer, Hermann (1994) *A Solar Manifesto*, James & James, London; Scheer, Hermann (2004) *The Solar Economy*, Earthscan, London, and Sterling, VA; Simmons, Matthew R. (2005) *Twilight in the Desert: The Coming Saudi Oil Shock and the World Economy*, Wiley, New York, NY; Stobaugh, Robert and Yergin, Daniel (1980) *Energy Future*, Ballantine Books, New York, NY; Tertzakian, Peter (2007) *A Thousand Barrels a Second: The Coming Oil Break Point and the Challenges Facing an Energy Dependent World*, McGraw-Hill, New York, NY; and Udall, Stewart, Conconi, Charles and Osterhaut, David (1975) *The Energy Balloon*, Penguin, New York, NY

5. Davis, Ged R. (1991) 'Energy for Planet Earth', *Readings from Scientific American*, W. H. Freeman and Company, New York, NY, p2

6. Campbell, Colin J. and Laherrère, Jean H. (1998)

7. Heinberg, Richard (2003); Klare, Michael T. (2002) *Resource Wars*, Henry Holt and Company, New York, NY; Klare, Michael T. (2004) *Blood and Oil: The Dangers and Consequences of America's Growing Petroleum Dependency*, Henry Holt and Company, New York, NY; and Yergin, Daniel (1992) *The Prize: The Epic Quest for Oil, Money & Power*, Simon & Schuster, New York, NY

8. Cappiello, Dina (2010) 'Toxic coal ash piling up in ponds in 32 states', Associated Press, www.usatoday.com/news/nation/environment/2009-01-09-coal-ash_N.htm, accessed 5 February 2010

9. Austen, Ian (2009) 'The Costly Compromises of Oil From Sand', *New York Times*, www.nytimes.com/2009/01/07/business/07oilsands.html?emc=etal, accessed 5 February 2010

10. *The Report of the UN's Intergovernmental Panel on Climate Change* (2007), www.ipcc.ch. Also, Global Tomorrow Coalition (1990) *The Global Ecology Handbook*, Beacon Press, Boston, MA; Gelbspan, Ross (1998) *The Heat Is On: The Climate Crisis, the Cover-Up, the Prescription*, Perseus, Boston, MA; Gelbspan, Ross (2004) *Boiling Point: How Politicians, Big Oil and Coal, Journalists and Activists are Fueling the Climate Crisis – and What We Can Do to Avert Disaster*, Basic Books, New York, NY; Gore, Al (1992) *Earth in the Balance*, Houghton Mifflin, Boston, MA; Gore, Al (2006) *An Inconvenient Truth*, Rodale Press, Emmaus, PA; Legett, Jeremy (2001) *Carbon War and the End of the Oil Era*, Routledge, New York; Lyman, Francesca, with Mintzer, Irving, Courrier, Kathleen and Mackenzie, James (1990) *The Greenhouse Trap*, Beacon Press, Boston, MA; McKibben, Bill (1989) *The End of Nature*, Random House, New York, NY; Monbiot, George (2007) *Heat: How to Stop the Planet from Burning*, South End Press, Cambridge, MA; Oppenheimer, Michael and Boyle, Robert H. (1990) *Dead Heat*, Basic Books, New York, NY; Schneider, Stephen and Londer, Randi (1984) *The Coevolution of Climate and Life*, Sierra Books, San Francisco, CA; Scientific American (1990) *Managing Planet Earth: Readings from Scientific American*, W. H. Freeman and Company, New York, NY; Speth, James Gustave (2004) *Red Sky at Morning*, Yale University Press, New Haven, CT, and London; Weart, Spencer R. (2003) *The Discovery of Global Warming*, Harvard University Press, Cambridge, MA

11. Global Tomorrow Coalition (1990), pp224–228; MacEachern, Diane (1990) *Save Our Planet*, Dell, New York, NY; Postel, Sandra (1985) 'Protecting Forests from Air Pollution and Acid Rain', in Lester R. Brown and others, *State of the World*, Norton, New York, NY

12. Bartlett, Donald and Steele, James (1985) *Forevermore: Nuclear Waste in America*, Norton, New York, NY; Berger (1977); Caldicott, Helen (1979) *Nuclear Madness*, Autumn Press, Brookline, MA; Caldicott, Helen (2002) *The New Nuclear Danger*, New Press, New York, NY; Caldicott, Helen (2006) *Nuclear Power Is Not the Answer*, New Press, New York, NY; Committee for

the Compilation on Damage Caused by the Atomic Bombing in Hiroshima and Nagasaki (1981) *Hiroshima and Nagasaki*, Basic Books, New York, NY; Browne, Corinne and Munroe, Robert (1981) *Time Bomb*, William Morrow, New York, NY; Calder, Nigel (1981) *Nuclear Nightmares*, Penguin, Harmondsworth, UK; Davenport, Elaine, Eddy, Paul and Gilman, Peter (1978) *The Plumbat Affair*, Futura Publications, London; Durie, Sheila and Edwards, Rob (1982) *Fuelling the Nuclear Arms Race*, Pluto Press, London; Environmental Action Foundation (1979) *Accidents Will Happen*, Harper & Row, New York, NY; Falk, Jim (1982) *Global Fission*, Oxford University Press, New York, NY; Faulkner, Peter (ed.) (1977) *The Silent Bomb*, Vintage Books, New York, NY; Ford, Daniel (1982) *Three Mile Island: Thirty Minutes to Meltdown*, Penguin, New York, NY; Ford, Daniel (1986) *Meltdown*, Simon & Schuster, New York, NY; Ford, Daniel, Kendall, Henry and Nadis, Steven (1982) *Beyond the Freeze*, Beacon Press, Boston, MA; Fuller, John (1975) *We Almost Lost Detroit*, Reader's Digest Press, New York, NY; Gale, Robert and Hauser (1988) Thomas, *Final Warning: The Legacy of Chernobyl*, Warner Books, New York, NY; Gofman, John and Tamplin, Arthur (1979) *Poisoned Power*, Rodale Press, Emmaus, PA; Gray, Mike and Rosen, Ira (1982) *The Warning*, Contemporary Books, Chicago, IL; Grossman, Karl (1980) *Cover Up: What You Are Not Supposed to Know About Nuclear Power*, Permanent Press, Segaponack, NY; Gyorgy, Anna and Friends (1979) *No Nukes*, South End Press, Boston, MA; Hawkes, Nigel and others (1987) *Chernobyl: The End of the Nuclear Dream*, Vintage Books, New York, NY; Hilgartner, Stephen, Bell, Richard and O'Connor, Rory (1983) *Nukespeak*, Penguin, Harmondsworth, UK; International Physicians for the Prevention of Nuclear War (IPPNW) and the Institute for Energy and Environmental Research (IEER) (1992) *Plutonium: Deadly Gold of the Nuclear Age*, International Physicians Press, Cambridge, MA; Kallet, Arthur and Schlink, F. J. (1935) *100,000,000 Guinea Pigs*, Grosset & Dunlap, New York, NY; Jungk, Robert (1979) *The New Tyranny*, Warner, New York, NY; League of Woman Voters Education Fund (1985) *The Nuclear Waste Primer*, Nick Lyons, New York, NY; Loeb, Paul (1982) *Nuclear Culture*, Coward, McCann & Geoghegan, New York, NY; Lovins, Amory and Lovins, Hunter (1981) *Energy/War: Breaking the Nuclear Link*, Harper & Row, New York, NY; Munson, Richard (ed.) (1978), *Countdown to a Nuclear Moratorium*, Environmental Action Foundation, Washington, DC; McKinley, Olson (1976) *Unacceptable Risk*, Bantam, New York, NY; Medvedev, Zhores (1980) *Nuclear Disaster in the Urals*, Vintage, New York, NY; Nader, Ralph and Abbots, John (1979) *The Menace of Atomic Energy*, Norton, New York, NY; Nelkin, Dorothy and Pollak, Michael (1982) *The Atom Besieged: Antinuclear Movements in France and Germany*, MIT Press, Cambridge, MA; Radical Science Collective (1984) *No Clear Reason*, Free Association Books, London; Rapoport, Roger (1971) *The Great American Bomb Machine*, Dutton, New York, NY; Reader, Mark (ed. (1980) *Atom's Eve*, McGraw-Hill, New York, NY; Roberts, Alan and Medvedev, Zhores (1977) *Hazards of Nuclear Power*, Spokesman, Nottingham, UK; Rosenberg, Howard (1980) *Atomic Soldiers: American Victims of Nuclear Experiments*, Beacon Press, Boston, MA; Safer, Thomas H. and Orville, Kelly E. (1983) *GI Victims of US Atomic Testing*, Penguin, New York, NY; Schell, Jonathan (1982) *The Fate of the Earth*, Alfred Knopf, New York, NY; Sternglass, Ernest (1981) *Secret Fallout*, McGraw-Hill, New York, NY; and Shapiro, Fred (1981) *Radwaste*, Random House, New York, NY

13. Nocera, Daniel, quoted in Anne Trafton, '"Major discovery" from MIT primed to unleash solar revolution', www.web.mit.edu/newsoffice/2008/oxygen-0731.html, accessed 5 February 2010

14. President Barack Obama, in his speeches on the economy on 22 January 2009 at George Mason University in Fairfax, VA, and on 24 January 2009 on the airwaves.

Chapter 1 The Sun: The Enduring Light

1. Shakespeare, William (1601) *Twelfth Night*, Act III, Scene 1
2. Tagore, Rabindranath (1966) *The Religion of Man*, Beacon Press, Boston, MA, p13
3. For some particularly illuminating discussions on the philosophical, spiritual, scientific and cultural significance and nature of the Sun see, Cole, John N. (1981) *Sun Reflections*, Rodale Press, Emmaus, PA; Gribbin, John (1991) *Blinded by the Light: New Theories about the Sun and the Search for Dark Matter*, Harmony Books, New York, NY; Jastrow, Robert (1980) *Until the Sun Dies*, Warner, New York, NY; Singh, Madanjeet (1993) *The Sun: Symbol of Power and Life*, Harry N. Abrams, Inc. and UNESCO, New York, NY; Singh, Madanjeet and others (1998) *The Timeless Energy of the Sun*, Sierra Club Books, San Francisco, CA; and Washburn, Mark (1981) *In the Light of the Sun*, Harcourt Brace Jovanovich, New York, NY, and London
4. Commoner, Barry (1990) *Making Peace with the Planet*, Pantheon, New York, NY, p148. Also see: Brower, Michael (1992) *Cool Energy: The Renewable Solutions to Environmental Problems*, MIT Press, Cambridge, MA; Daniels, Farrington (1977) *Direct Use of the Sun's Energy*, Ballantine, New York, NY; Droege, Peter (ed.) (2009) *100% Renewable: Energy Autonomy in Action*, Earthscan, London and Sterling, VA; Flavin, Christopher and Lenssen, Nicholas (1994) *Power Surge*, Norton, New York, NY; Gribbin, John (1991); Halacy, D. S. (1975) *The Coming Age of Solar Energy*, Avon Books, New York, NY; Hayes, Denis (1977) *Rays of Hope*, Norton, New York, NY; Hoffman, Jane and Hoffman, Michael (2008) *Green: Your Place in the New Energy Revolution*, Palgrave Macmillan, New York, NY; Lovins, Amory (1979) *Soft Energy Paths*, Harper & Row, New York, NY; Scheer, Hermann (1994) *A Solar Manifesto*, James & James Ltd, London; Scheer, Hermann (2004) *The Solar Economy*, Earthscan, London and Sterling, VA; Washburn, Mark (1981); Williams, Robert H. (ed.) (1978) *Towards a Solar Civilization*, MIT Press, Cambridge, MA; and Vaitheeswaran, Vijay V. (2005) *Power to the People*, Farrar, Straus and Giroux, New York, NY
5. Nocera, Daniel, quoted in Trafton, Anne, '"Major discovery" from MIT primed to unleash solar revolution', www.web.mit.edu/newsoffice/2008/oxygen-0731.html, accessed 5 February 2010
6. www.pioneers-of-power.de and 'Action-packed solar thriller' (2006), *Photon International*, July, p68
7. www.en.wikipedia.org/wiki/Global_dimming, accessed 5 February 2010
8. Wines, Michael (2009) 'Beijing's Air Is Cleaner, But Far From Clean', *New York Times*, 17 October
9. Bradshier, Keith (2009), 'China Vies to Be World's Leader in Electric Cars', *New York Times*, 1 April 2009, www.nytimes.com/2009/04/02/business/global/02electric.html?emc=eta1&pagewanted=print, accessed 5 February 2010

Chapter 2 Power to the People: Renewable Energy Technologies – Now!

1. Gandhi, M. K. (1967) *The Wisdom of Gandhi*, Philosophical Library, New York, NY, p46
2. Lovins, Amory B. (1979) *Soft Energy Paths*, Harper & Row, New York, NY, p24
3. Davidson, Joel (1989) *The New Solar Electric Home*, AATEC, Ann Arbor, MI, p1. Also: Fowler, Paul Jeffrey (1995) *The Evolution of an Independent Home: The Story of a Solar Electric Pioneer*,

Fowler Enterprises, Worthington, MA; Fowler Solar Electric Inc. (1991) *The Solar Electric Independent Home Book*, Fowler Solar Electric Inc., Worthington, MA; Halacy, Dan (1984) *Home Energy*, Rodale Press, Emmaus, PA; Hankins, Mark (1993) *Solar Rural Electrification in the Developing World* (Four country case studies: Dominican Republic, Kenya, Sri Lanka and Zimbabwe), Solar Electric Light Fund, Washington, DC; Kamal, Sajed (1989) *Photovoltaics: A Global Revolution and Its Scope for Bangladesh*, UBINIG – Policy Research for Development Alternative, Dhaka, Bangladesh; Maycock, Paul D. and Stirewalt, Edward N. (1985) *A Guide to the Photovoltaic Revolution*, Rodale Press, Emmaus, PA; Potts, Michael (1993) *The Independent Home*, Chelsea Green Publishing, Post Mills, VT; Starr, Gary (1987) *The Solar Electric Book*, Integral Publishing, in association with Solar Electric, Lower Lake, CA; and Strong, Steven J. with Scheller, William G. (1991) *The Solar Electric House*, Sustainability Press, Still River, MA

4. Komoto, Keiichi, Masakazu, Ito, Vleuten, Peter van der, Faiman, David and Kurokawa, Kosuko (eds) (2009) *Energy from the Desert: Very Large Photovoltaic System: Socio-economic, Financial, Technical and Environmental Aspects*, Earthscan, London and Sterling, VA. Also Bradsher, Keith (2009) 'Green Power Takes Root in the Chinese Desert', *New York Times*, 2 July 2009, www.nytimes.com/2009/07/03/business/energy-environment/03renew.html?pagewanted=2&_r=1, accessed 17 February 17 2010

5. www.news.discovery.com/space/japan-solar-space-station.html, accessed 9 February 2010

6. Ashley, Steven (1989) 'New Life for Solar?', *Popular Science*, May

7. www.us.sunpowercorp.com/utility/success-stories/german-solarpark-exceeds-energy-expectations.php, accessed 5 February 2010

8. www.solarserver.de/solarmagazin/anlage_0606_e.html, accessed 5 February 2010

9. www.us.sunpowercorp.com/utility/success-stories/serpa-power-plant-produces-consistent-energy-roi.php, accessed 5 February 2010

10. www.pvresources.com/en/top1000pv.php, accessed 6 February 2010

11. Naar, Jon (1990) *Design for a Livable Planet*, Harper & Row, New York, NY, p198

12. Naar, Jon (1982) *The New Wind Power*, Penguin, New York, NY, p71. Also: Gipe, Paul (1993) *Wind Power for Home and Business*, Chelsea Green Press, Post Mills, VT

13. Naar, Jon (1982), p199

14. Brower, Michael (1990) *Cool Energy*, Union of Concerned Scientists, Cambridge, MA, p45

15. 'Air Power' (2001), *NBC Nightly News*, 17 May; and '21 Frequently Asked Questions About Wind Energy', Danish Wind Turbine Manufacturers Association, www.windpower.dk/faqs.htm

16. Howe, Peter J. (2008) 'Wind turbines propel Logan's energy efforts', www.boston.com/news/local/articles/2008/03/05/wind_turbines_propel_logans_energy_efforts, accessed 5 February 2010

17. *The European Offshore Wind Industry – Key Trends and Statistics 2009* (2009), www.windpower.org/download/450/Offshore_statistik_2009.PDF, accessed 9 February 2010

18. www.capewind.org; Williams, Wendy and Whitcomb, Robert (2008) *Cape Wind: Money, Celebrity, Energy, Class, Politics, and the Battle for Our Energy Future*, Public Affairs, New York, NY

19. www.energy-revolutions.com

20. www.gwec.net

21. Rackstraw, Kevin (1998) 'Wind Around the World', *Solar Today*, March/April; and Naar, Jon (1982), p200

22. Naar, Jon (1990), p196

23. Brower, Michael (1990), p70

24. Naar, Jon (1990), p197

25. Jackson, David S. (1988) 'Aswan, Drought Stalks the Mighty Nile', *Time*, 28 March, p79

26. Switkes, Glenn, 'Power-less and Damned', www.citizen.org/cmep/Water/cmep_Water/reports/brazil/articles.cfm?ID=8748, accessed 6 February 2010

27. McCully, Patrick (2001) *Silenced Rivers: The Ecology and Politics of Large Dams*, Zed Books, London; Morgan, Arthur E. (1971) *Dams and Other Disasters: A Century of Army Corps of Engineers in Civil Works*, Porter Sargent Publisher, Boston, MA; Roy, Arundhati (2002) *Power Politics*, South End Press, Cambridge, MA; and World Commission on Dams (ed.) (2001) *Dams and Development: A New Framework for Decision-Making*, Earthscan, London. Also see: www.narmada.org and Allen, Bruce (2008) *Lake of Heaven: An original translation of the Japanese novel by Ishimure Michiko*, Lexington, Lanham, MD (a fictional work that is rooted in the history of actual mountain villages in Japan destroyed in the process of constructing a dam)

28. Chow, Elaine (2009) 'Massive Hydroelectric Dams Could Have Caused the Sichuan Earthquake', www.gizmodo.com.au/2009/02/massive_hydroelectric_dams_could_have_caused_the_sichuan_earthquake-2/, accessed 6 February 2010; Ibotombi, Soibam (2009) 'Tipaimukh Dam is a Geo-tectonic Blunder of International Dimensions', http://inpui.blogspot.com/2009/05/paper-tipaimukh-dam-is-geo-tectonic.html, accessed 6 February 2010

29. Davis, Scott (ed.) (2010) *Serious Microhydro*, New Society Publishers, Gabriola Island, BC, Canada. For a selection of microhydros around the world, access www.energy.sourceguides.com

30. Jackson, Onaje (2000) 'Investing in Sustainability at Coral World', *Solar Today*, July/August

31. Singh, Dilawar (1994) 'Water pumping and electricity from a river flow turbine', *SunWorld*, June, p13

32. Norman, Colin (1981) *The God That Limps*, Norton, New York, NY, p170

33. Bakthavatsalam, Dr V. (2001) 'Windows of Opportunity: IREDA and the role of RE in India', *REFOCUS*, May

34. Butti, Ken and Perlin, John (1980) *A Golden Thread: 2500 Years of Solar Architecture and Technology*, Cheshire, Palo Alto, CA, pp117–119

35. Schaeffer, John & the Real Goods Staff (2007), *Solar Living Sourcebook*, Chelsea Green Publishing Company, White River Junction, VT

36. Berry, Wendell (1978) *The Unsettling of America*, Avon, New York, NY; Carson, Rachel (1964) *Silent Spring*, Fawcett Crest, New York, NY; Doyle, Jack (1985) *Altered Harvest*, Viking, New York, NY; Everest, Larry (1985) *Behind the Poison Cloud: Union Carbide's Bhopal Massacre*, Banner Press, Chicago, IL; Fox, Dr Michael W. (1986) *Agricide*, Schocken, New York, NY; Howard, Sir Albert (1975) *The Soil and Health*, Schocken, New York, NY; Jackson, Wes, Berry, Wendell and Colman, Bruce (eds) (1984) *Meeting the Expectations of the Land*, North Point Press, San Francisco, CA; Lappe, Frances Moore and Collins, Joseph (1977) *Food First: Beyond the Myth of Scarcity*, Houghton Mifflin, Boston, MA; Merrill, Richard (ed.) (1976), *Radical Agriculture*, Harper & Row, New York, NY; Mott, Lawrie and Snyder, Karen (1988) *Pesticide Alert*, Sierra Club Books, San Francisco, CA; Regenstein, Lewis (1982) *America the Poisoned*, Acropolis, Washington, DC; and Whiteside, Thomas (1970) *Defoliation*, Ballantine, New York, NY

37. Edey, Anna (1998) *Solviva*, Trailblazer Press, Martha's Vineyard, MA; Anna Edey (1991) 'The Twelve-month Growing Season in All Climates', in Paul Rothkrug and Robert L. Olson (eds) *Mending the Earth*, North Atlantic, Berkeley, CA and Environmental Rescue Fund, San Francisco, CA; and Perlmutter, Cathy (1989) 'Greenhouse Gold', *Organic Gardening*, October

38. www.badgersettfarm.com. Also: 'Badgersett Research Farm Greenhouse' (1993) *Real Goods News*, October; Gundersen, Roald (undated) 'The Solar Garden: Solar Greenhouses and Sustainable Living', www.mwt.net/~roald/llewellyn.html, accessed 6 February 2010; Gunderson, Roald (undated) 'Solar Cold Climate Greenhouse (SCCG)', www.mwt.net/~roald/solargh.html, accessed 6 February 2010; and 'Southeast Clean Energy Resource Team's Badgersett Research Farm Tour Summary, Amherst, MN, October 31, 2007' (2007), www.cleanenergyresourceteams. org/files/SESummary10-31-07.pdf, accessed 6 February 2010

39. Hossain, Rokeya Sakhawat (1988) *Sultana's Dream and Selections from The Secluded Ones*, Feminist Press at City University of New York, NY; Rokeya, Begum (1993) *Rokeya Rachanabali* (collected works of Begum Rokeya), Bangla Academy, Dhaka, Bangladesh; and Rokeya, Begum (1998) *Sultana's Dream* (booklet), Narigrantha Prabartana, Dhaka, Bangladesh

40. For a brief history of solar cooker technology, see Butti, Ken and Perlin, John (1980); and Halacy, Beth and Dan (1992) *Cooking with the Sun*, Morning Sun Press, Lafayette, CA (this book also includes instructions on how to build and use several types of solar cookers)

41. Halacy, Beth and Dan (1992), p41

42. Search for 'Tulsi-Hybrid Solar Cooker Oven' on the internet for several informational and distribution sources.

43. Graham, James (2001) 'Ripening RE Markets: Capacity Building for the Rapid Commercialization of RE in China', *REFOCUS*, April

44. 'One in Every Ten Guangxi Farmhouses Uses Biogas' (2000), *People's Daily*, 17 July, www.english. peopledaily.com.cn/200007/17/eng20000717_45675.html, accessed 6 February 2010

45. Bakthavatsalam, Dr V. (2001) 'Windows of Opportunity: IREDA and the role of RE in India', *REFOCUS*, May

46. www.arti-india.org

47. Wheeler, Patrick (2000) 'Commercial and Strategic Perspectives for Anaerobic Digestion', *International Directory of Solid Waste Management 2000/2001 ISWA Yearbook*; 'The AgSTAR Program', US Environmental Protection Agency, www.epa.gov/agstar/library/script.html and Texas A&M University-Biogas Project

48. Hoffmann, Peter (2002) *Tomorrow's Energy: Hydrogen, Fuel Cells, and the Prospects for a Cleaner Planet*, MIT Press, Cambridge, MA; Ogden, Joan M. and Williams, Robert H. (1989) *Solar Hydrogen: Moving Beyond Fossil Fuels*, World Resource Institute, Washington, DC; Lehman, Peter and Parra, Christine (1994) 'Hydrogen Fuel From the Sun', *Solar Today*, September/ October; Peavey, Michael A. (1993) *Fuel From Water*, Merit Products, Inc., Louisville, KY; Rifkin, Jeremy (2003) *The Hydrogen Economy*, Tarcher/Penguin, New York, NY; and Romm, Joseph J. (2004) *The Hype About Hydrogen*, Island Press, Washington, DC. Also: www.merit.unu. edu/publications/pb/unu_pb_2006_03.pdf, accessed 6 February 2010

49. Nocera, Daniel (2008) http://web.mit.edu/newsoffice/2008/oxygen-0731.html, accessed 6 February 2010

50. Nocera, Daniel (2008) 'Powering the Planet: The Challenge for MIT and Science in the 21st Century', public lecture at MIT, 15 September

51. Monahan, Patricia (2008) 'Biofuels: An Important Part of a Low-Carbon Diet', *Catalyst*, Spring; Werner, Carol (2008) 'Resolving the Biofuels Dilemma', *Solar Today*, July/August; and Grunwald, Michael (2008) 'The Clean Energy Scam', *Time*, 7 April

52. Quoted in Hewitt, Bill (2010) 'Algae', www.climatechange.foreignpolicyblogs.com/?s=algae, accessed 6 February 2010

Chapter 4 **From the Collapsing Economy to a Sustainable Economy: The Real Economic Advantages of Renewable Energy Technologies**

1. Schumacher, E. F. (1977) *A Guide for the Perplexed*, Harper & Row, New York, NY, p140

2. Goodstein, David (2004) *Out of Gas: The End of the Age of Oil*, Norton, New York, NY, p123. Also: Diamond, Jared (2005) *Collapse: How Societies Choose to Fail or Succeed*, Penguin, Harmonsdworth; and Ponting, Clive (2007) *A New Green History of the World: The Environment and the Collapse of Great Civilizations*, Penguin, Harmondsworth

3. Nicklas, Mike (1989) 'Americans Want Environmentally Sound Energy Options', *Solar Today*, September/October. Also: Brower, Michael (1992) *Cool Energy: Renewable Solutions to Environmental Problems*, MIT Press, Cambridge, MA; Hubbard, Harold M. (1991) 'The Real Cost of Energy', *Scientific American*, April; and Larson, R., Vignola, F., and West, R. (eds) (1992) *Economies of Solar Energy Technologies*, American Solar Energy Society, Boulder, CO

4. Silverthorne, Katherine (1998) *The Rising Cost of Global Warming*, US Public Interest Research Group, Washington, DC, p2

5. UNEP (2001) 'Impact of Climate Changes To Cost The World $US 300 Billion A Year', United Nations Environment Programme, Press Release, February, www.unep.org/Documents/Default.asp?DocumentID=192&ArticleID=2758, accessed 6 February 2010

6. Waldman, Peter (1992) 'Echoes From Gulf War Still Reverberate And Encumber King Hussein of Jordan', *Wall Street Journal*, 9 June

7. Hubbard, Harold M. (1991). Also see, DePaola, Ross (1991) 'Don't Get Me Started on the War in Iraq', *SunWorld*, March/April, p3

8. www.costofwar.com

9. Reuters (2007) 'US CBO estimates $2.4 trillion long-term war costs', 24 October, www.reuters.com/article/idUSN2450753720071024, accessed 6 February 2010

10. Stiglitz, Joseph (2008) *The Three Trillion Dollar War*, Norton, New York

11. Tamminen, Terry (2006) *Lives Per Gallon: The True Cost of Our Oil Addiction*, Island Press, Washington, DC

12. Krosinsky, Cary and Robins, Nick (2008) *Sustainable Investing: The Art of Long-Term Performance*, Earthscan, London and Sterling, VA

13. World Commission on Environment and Development (1987), *Our Common Future*, Oxford University Press, Oxford, UK, and New York, NY, pp28–29

14. Daly, Herman E. and Cobb, Jr., John B. (1989) *For the Common Good: Redirecting the Economy Toward Community, the Environment, and a Sustainable Future*, Beacon Press, Boston, MA, p21. Also: Bartelmus, Peter (1986) *Environment and Development*, Allen & Unwin, Boston, MA; Commoner, Barry (1976) *The Poverty of Power: Energy and the Economic Crisis*, Alfred Knopf, New York, NY; Daly, Herman and Farley, Joshua (2003) *Ecological Economics: Principles and Applications*, Island Press, Washington, DC; Heilbroner, Robert L. (1975) *An Inquiry into the Human Prospect*, Norton, New York; Heilbroner, Robert L. and Thurow, Lester (1981) *Five Economic Challenges*, Prentice-Hall, Englewood Cliffs, NJ; Henderson, Hazel (1980) *Creating Alternative Futures*, G. P. Putnam's Sons, New York, NY; Korten, David C. (1995) *When*

Corporations Rule the World, Kumarian Press (with Berret-Koehlar Publishers), West Hartford, CT; Martinez-Alier, Juan with Schlupmann, Klaus (1987) *Ecological Economics*, Blackwell, Oxford, UK, and Cambridge, MA; Nadeau, Robert L. (2006) *The Environmental Endgame: Mainstream Economics, Ecological Disaster, and Human Survival*, Rutgers University Press, Piscataway, NJ; Rahnema, Majid with Bowtree, Victoria (1997) *The Post-Development Reader*, Zed Books, London; Schumacher, E. F. (1973) *Small Is Beautiful*, Harper & Row, New York; and Peet, John (1992) *Energy and Ecological Economics of Sustainability*, Island Press, Washington, DC

15. Pearce, David, Markandya, Anil and Barbier, Edward (1989) *Blueprint for a Green Economy*, Earthscan, London, pp5–7

16. Pearce, David, Markandya, Anil and Barbier, Edward (1989), p30

17. Andrejko, Dennis A. (ed.) (1989) *Assessment of Solar Energy Technologies*, American Solar Energy Society, Boulder, CO; Hubbard, H. M., Notari, P., Deb, S. and Awerbuch, Leon(1994) *Progress in Solar Energy Technologies and Applications*, American Solar Energy Society, Boulder, CO; Brower, Michael (1992); Union of Concerned Scientists (1992) 'Environmental Impacts of Renewable Energy Technologies', Union of Concerned Scientists, Cambridge, MA, August; and Larson, R., Vignola, F., and West, R. (eds) (1992)

18. www.unep.org/pdf/PressReleases/Reforming_Energy_Subsidies.pdf, accessed 6 February 2010

19. Hansen, Richard and Martin, Jose (1988) 'Photovoltaics for Rural Development in the Dominican Republic', in United Nations, *Natural Resources Forum*, Graham & Trotman, London, pp115–128; Covell, Philip (1990) 'A Bright Idea: Photovoltaics in the Dominican Republic', *Solar Today*, March/April; *SunWorld* (1990) 'Indonesia to Install 2000 Solar Energy Systems', September; Kamal, Sajed (1991) 'The Photovoltaic Revolution: A Global Revolution & Its Scope for Bangladesh', ICED Report, International Consortium for Energy Development (ICED), Boston, MA; Kamal, Sajed (2008) 'The Untapped Energy Mine: The Revolutionary Scope of Renewable Energy for Bangladesh', *Journal of Bangladesh Studies*, vol 10, no 1, www.bdiusa.org; and Mondal, M., Alam Hossain and Chakrabarty, Sayan (2009) 'Socio-Economic Impacts of the Solar Home Systems in Bangladesh', a paper presented at the Conference on Ideas and Innovations for the Development of Bangladesh: The Next Decade, John F. Kennedy School of Government, Harvard University, 9–10 October

20. Serchuk, Dr Adam (2000) 'The Environmental Imperative for Renewable Energy', *Solar Today*, November/December

21. Kyocera's New Headquarters Building, Fushimi-ku, Kyoto, www.//arch.hku.hk/~cmhui/japan/kyocera/kyocera-index.html, accessed 6 February 2010

22. Jürgens, J., Szacsvay, T., Brown, J., Eckmanns, A. and Posanansky, M. (1998) 'PV roofing materials – key to the large scale generation of solar electricity: experiences with SUNSLATES™', Paper for the Vienna PV world conference in July

23. www.spiresolar.com

24. Maycock, Paul D. and Stirewalt, Edward N. (1985) *A Guide to the Photovoltaic Revolution*, Rodale Press, Emmaus, PA , p71; and www.nytimes.com/ads/peoplesoft/article9

25. Morris, Craig (2006) *Energy Switch: Proven Solutions for a Renewable Future*, New Society Publisher, Gabriola Island, BC, Canada, p83. Also see www.solar-fabrik.de, accessed 5 February 2010.

26. For some incisive analyses of Third World debt and its consequences, see Hayter, Teresa and Watson, Catherine (1985) *Aid: Rhetoric and Reality*, Pluto Press, London; Hofrichter, Richard (ed.) (1993) *Toxic Struggles: The Theory and Practice of Environmental Justice*, especially the articles under the section 'The Global Connection: Exploitation of Developing Countries',

New Society Publishers, Philadelphia, PA; Rahnema, Majid with Bawtree, Victoria (1997); and Sobhan, Rehman (1982) *The Crisis of External Dependence*, University Press Limited, Dhaka, Bangladesh

27. BBC News (2007) 'Oil imports hike US trade deficit', 12 July, www.news.bbc.co.uk/2/hi/business/6895738.stm, accessed 6 February 2010

28. Calleo, David P. (1993) *The Bankrupting of America*, Avon Books, New York, 1993

29. www.warresisters.org, 'Where Your Income Tax Money Really Goes', United States Federal Budget for Fiscal Year 2003

30. Union of Concerned Scientists (1994) 'The Global Environmental Crisis: Causes, Connections, and Solutions', Publications Department BP, UCS, Cambridge, MA, p3

31. United Nations Framework Convention on Climate Change (2007) *Climate Change: Impacts, Vulnerabilities and Adaptation in Developing Countries*, p5, www.unfccc.int/files/essential_background/background_publications_htmlpdf/application/txt/pub_07_impacts.pdf, accessed 6 February 2010

32. Daily Star (2009) 'Maldives presses for action, not words', 19 September, www.thedailystar.net/newDesign/news-details.php?nid=106399, accessed 6 February 2010

33. McKibben, Bill (2009) 'Copenhagen: Thing Fall Apart and an Uncertain Future Looms', *Yale Environment 360*, 21 December, http://e360.yale.edu/content/feature.msp?id=2225, accessed 7 February 2010; and Huq, Saleemul (undated) 'From Hopenhagen to Brokenhagen: A Personal Journey', One World website, www.uk.oneworld.net/article/view/164255/1/?PrintableVersion=enabled, accessed 7 February 2010

34. Yergin, Daniel (1992) *The Prize: The Epic Quest for Oil, Money & Power*, Simon & Schuster, New York, NY. Also: Brisard, Jean-Charles and Dasque (2002) *Forbidden Truth: UK-Taliban Secret Oil Diplomacy and the Failed Hunt for Bin Laden*, Thunder's Mouth Press/Nation Books, New York, NY; Ghazvinian, John (2007) *Untapped: The Scramble for Africa's Oil*, Harcourt, New York, NY; Heinberg, Richard (2003) *The Party's Over: Oil, War and the Fate of Industrial Societies*, New Society, Gabriola Island, BC, Canada; Klare, Michael T. (2001) *Resource Wars: The New Landscape of Global Conflict*, Henry Holt and Company, New York, NY; Klare, Michael T. (2004) *Blood and Oil: The Dangers and Consequences of America's Growing Petroleum Dependency*, Henry Holt and Company, New York, NY; McQuaig, Linda (2006) *It's the Crude, Dude*, Thomas Dunne Books, St Martin's Press, New York, NY; Shaxson, Nicholas (2008) *Poisoned Wells: The Dirty Politics of African Oil*, Palgrave Macmillan, New York, NY; and Vidal, Gore (2002) *Dreaming War: Blood for Oil and the Cheney-Bush Junta*, Thunder's Mouth Press/Nation Books, New York, NY

35. Epstein, Paul R. and Achebe, Chidi (2003) 'Prize or Curse?', *Boston Globe*, 2 May 2003

36. Barnet, Richard J. (1981) *Real Security*, Simon and Schuster, New York, NY, p98

37. *Global Fissile Material Report 2007* of the International Panel on Fissile Materials, www.fissilematerials.org

38. Jungk, Robert (1979) *The New Tyranny*, Warner Books, New York, NY, p9

39. Lovins, Amory B. and Lovins, L. Hunter (1981) *Energy/War: Breaking the Nuclear Link*, Harper & Row, New York, NY. Also: www.rmi.org

40. Sengupta, Somini (2007) 'Glaciers in Retreat', *New York Times*, 17 July 2007

41. Eisenberg, Daniel (2000) 'Power's Surge', *Time*, 17 July 2000, p37

42. Navigant Consulting (2004) *The Changing Face of Renewable Energy*, Navigant Consulting, Burlington, MA, p8, www.navigantconsulting.com

43. Bradford, Travis (2006) *Solar Revolution: The Economic Transformation of the Global Energy Industry*, MIT Press, Cambridge, MA. An investment banker, Travis Bradford offers an excellent

analysis of the lowering trend of energy generated by photovoltaics and projects meaning that it will soon become cheaper than any other energy source. Also, Robertson, Jordan (2010) 'Where's the next boom? May be "cleantech"', *ABC News*, www.abcnews.go.com/Technology/ wirestory?id=8766490&page=4, accessed 17 February 2010, and Takahashi, Dean (2010) 'Solar Boom', *Technology Review*, September/October, www.technologyreview.com/Energy/21228/, accessed 17 February 2010

44. Laird, Frank N. (2003) 'Fighting Over Crumbs', *Solar Today*, November/December 2003, p18
45. Lagatta, Daniel P. (2010) GEI Consultants Press Release, www.geiconsultants.com/nuclear-power--a-green-technology--pdf, accessed 17 February 2010
46. Myers, Norman (1985) *Gaia: An Atlas of Planet Management*, Doubleday, New York, NY; Badiner, Allan Hunt (ed.) (1990) *Dharma Gaia*, Parallax Press, Berkeley, CA; and Harding, Stephen (2006) *Animate Earth: Science, Intuition, and Gaia*, Chelsea Green, White River Junction, VT
47. World Wildlife Fund (2006) *Living Planet Report 2006*, WWF, Gland, Switzerland, p1, www.panda.org/livingplanet
48. Gore, Al (2006) *An Inconvenient Truth: The Planetary Emergency of Global Warming and What We Can Do About It*, Rodale Press, New York

Chapter 5 The Renewable Revolution: Turning Vision into Action

1. Houch, Kristie Sue (1971), in *What Are Me and You Gonna Do? Children's Letters to Senator Gaylord Nelson about the Environment*, Ballantine, New York, NY
2. Ban Ki-moon (2009) www.earthtimes.org/articles/show/280809,un-chief-calls-for-global-push-to-combat-climate-change.html, accessed 6 February 2010
3. Dr Seuss (1971) *The Lorax*, Random House, New York, NY
4. Halacy, D. S. (1975) *The Coming Age of Solar Energy*, Avon Books, New York, NY, p237
5. Hayes, Denis (1977) *Rays of Hope*, Norton, New York, NY, p25. Also see Capra, Fritjof (1988) *The Turning Point*, Bantam, New York, NY, p408
6. Lovins, Amory (1993), foreword to *Real Goods Book of Lights*, Real Goods, Ukiah, CA. Also: Ficket, Arnold, Gellings, Clark and Lovins, Amory (1991) 'Efficient Use of Electricity', *Energy for Planet Earth: Readings from Scientific American*, W. H. Freeman, New York, NY; Flavin, Christopher and Durning, Alan (1988) 'Raising Energy Efficiency', in Lester Brown and others, *State of the World 1988*, Norton, New York, NY; and Gibbons, John, Blair, Peter and Gwin, Holly (1989) 'Strategies for Energy Use', *Scientific American*, September
7. www.zipcar.org
8. Grossman, Elizabeth (2006) *High Tech Trash: Digital Devices, Hidden Toxics, and Human Health*, Island Press, Washington, DC
9. Lovins, Amory and Hunter (1991) 'Make Fuel Efficiency Our Gulf Strategy', *Alternative Energy Sourcebook*, Real Goods, Ukiah, CA
10. Thottam, Jiyoti (2009) 'Nano Power', *Time*, 13 April 2009
11. Illich, Ivan (1974) *Energy and Equity*, Harper & Row, New York, NY
12. Kay, Jane Holtz (1998) *Asphalt Nation: How the Automobile Took Over America, and How We Can Take It Back*, University of California Press, Berkeley, CA
13. Quoted in Southworth, Michael and Ben-Joseph, Eran (1997) *Streets and the Shaping of Towns and Cities*, McGraw-Hill, New York, NY, p53

14. www.earthhour.org

15. Earthworks Group (1989) *50 Simple Things You Can Do To Save The Earth*, Earthworks, Berkeley, CA

16. Earthworks Group (1990) *50 Simple Things Kids Can Do To Save The Earth*, Andrews and McMeel, Kansas City, MO, and New York, NY

17. MacEachern, Diane (1990) *Save Our Planet: 750 Everyday Ways You Can Help Clean Up the Earth*, Dell, New York, NY

18. Naar, Jon (1990) *Design for a Livable Planet: How You Can Help Clean Up the Environment*, Harper & Row, New York, NY

19. Brower, Michael and Leon, Warren (1999) *Consumer's Guide to Effective Environmental Choices*, Crown, New York, NY

20. Rogers, Elizabeth and Kostigen, Thomas M. (2007) *The Green Book: The Everyday Guide to Saving the Planet One Simple Step at a Time*, Three Rivers Press, New York

21. Asmus, Peter (1998) 'Power to the People – The Promise of Green Municipal Aggregation', *Solar Today*, May/June, p26

22. Asmus, Peter (1998), p31

23. Howe, Peter J. (2008), 'Wind turbines propel Logan's energy efforts', *Boston Globe*, 5 March, www.boston.com/news/local/articles/2008/03/05/wind_turbines_propel_logans_energy_efforts, accessed 6 February 2010

24. www.solardesign.com

25. Strong, Steven J. (1987) *The Solar Electric House*. Originally published in 1987 by Rodale Press in 1987, the book is now available through www.solardesign.com/book

26. www.irecusa.org/schools/index.html

27. www.massenergy.com

28. www.cityofboston.gov/climate/solar.asp

29. www.soltrex.com/systems.cfm?systemid=S00000000258, accessed 6 February 2010

30. www.genasun.com

31. www.boston.redsox.mlb.com/content/printer_friendly/bos/y2008/m05/d19/c2730414.jsp, accessed 6 February 2010

32. Ross, Casey (2009) 'Big solar statement for Fenway Center', *Boston Globe*, 4 December, www.boston.com/realestate/news/articles/2009/12/04/big_solar_statement_for_fenway_center, accessed February 21, 2010; and telephone conversation with Jerry Belair, Project Manager and Director of Leasing, Meredith Management Corporation, www.meredithmanagement.com, 6 October 2009

33. www.heat.neu.edu

34. www.solarfenway.org

35. www.solarhouse.com and www.solardesign.com/projects/project_display.php?id=30

36. www.spirecorp.com

37. Radcliffe, David (2006), quoted in *USA Today*, 17 October 2006

38. www.heliotronics.com

39. www.hullwind.org

40. Desai, Pooran and King, Paul (2006) *One Planet Living*, Alastair Sawday, Bristol, UK. Also, www.oneplanetliving.org

41. www.young-germany.de/life-in-germany/life-in-germany/article/6acb89b236/freiburg-germanys-green, accessed 6 February 2010

42. www.transatlantic21.org

43. www.martinvosseler.ch/sites/sunwalk/s1e.htm, accessed 6 February 2010

44. www.consumerenergycenter.org and www.gosolarcalifornia.ca.gov/about.html, accessed 6 February 2010

45. www.mtpc.org/massrenew

46. Hansen, Richard and Martin, Jose (1988) 'Photovoltaic for Rural Development in the Dominican Republic', *Natural Resources Forum*, published for the United Nations by Graham & Trotman, London, May, pp115–128; and Covell, Philip (1990) 'A Bright Idea: Photovoltaics in the Dominican Republic', *Solar Today*, March/April

47. McDowell, Fiona (2003) 'Rural Energy in Bangladesh', *Solar Today*, March/April. Also see www.gshakti.org

48. Kamal, Sajed (2008) 'The Untapped Energy Mine: The Revolutionary Scope of Renewable Energy in Bangladesh', *Journal of Bangladesh Studies*, vol 10, no 1, www.bdiusa.org. Also, www.brac.net/index.php?nid=88

49. www.rahimafrooz.com

50. 'Focus on India' (2001) includes four articles by different authors which provide an excellent summary of the Indian renewable energy programme, *Refocus*, May, www.re-focus.net. Also: Ringwald, Alexis (2009) 'India Charts Course for Renewable Energy Superpower', *Solar Today*, April

51. www.centerforfinancialinclusion.org/Page.aspx?pid=1392, accessed 6 February 2010

52. www.reuters.com/article/pressRelease/idUS142875+11-Nov-2008+BW20081111, accessed 6 February 2010

53. www.wholefoods.com

54. www.usgbc.org/LEED

55. Masia, Seth (2009) 'Back to the Future', *Solar Today*, March. Also, www.solartoday.org/nextwest

56. www.hmfh.com

57. www.whrc.org

58. Wilson, Alex (2003) 'Green Building – A Natural for This Biological Preserve', *Solar Today*, September/October

59. www.nesea.org

60. www.thehoya.com/node/5226, accessed 6 February 2010

61. Sullivan, C.C. (2008) 'Shades of Solar', Buildings, September, www.buildings.com/DesktopModules/BB_ArticleMax/ArticleDetail/BBArticleDetailPrint.aspx?ArticleID=6510&Template=Standard_Print.ascx&siteID=1, accessed 6 February 2010

62. www.nesea.org

63. www.solarnow.org

64. www.ssl.mit.edu/nasa_epo/solar, accessed 6 February 2010

65. Mazumder, Rezaul Karim (2008) 'Grid-connected solar power shines at DU campus', *Daily Star*, 28 October, www.thedailystar.net/pf_story.php?nid=60011, accessed 6 February 2010

66. A 'Climate Change Information and Strategy Packet' is available without cost by calling 1-888-9climate. It includes 'It's God's World: Christians, Care for Creation and Global Warming', a five-session Bible study for congregations

67. Partridge, Catherine (2002) 'Interfaith Power and Light urges diocese to use its energy well', *Episcopal Times*, January, p4. Also, www.MIPandL.org and www.theregenerationproject.org

68. Mazumder, Rezaul Karim (2008)

69. Motavalli, Jim (2002) 'Stewards of the Earth', *E Magazine*, November/December

70. 'Nuclear power phase-out', www.en.wikipedia.org/wiki/Nuclear_power_phase-out

71. Aiken, Donald, 'Germany Launches Its Transition to All Renewables', www.sustainablebusiness. com/features/feature_printable.cfm?ID=1208

72. www.wind-works.org/FeedLaws/HermannScheerLessonsfromGermany.html, accessed 6 February 2010

73. Runci, Paul (2005) 'Renewable Energy Policy in Germany: An Overview and Assessment', a report by the Joint Global Change Research Institute, University of Maryland, College Park, Maryland, January. Also, Morris, Craig (2006) *Energy Switch: Proven Solutions for a Renewable Energy Future*, New Society Publishers, Gabriola Island, BC, Canada

74. *The Power of Community: How Cuba Survived Peak Oil*, DVD produced by the Community Service, Inc., P.O. Box 243, Yellow Springs, Ohio 45387. Available through www.communitysolution. org/cuba

75. Environment News Service (ENS) (2008), 17 July, www.ens-newswire.com/ens/jul2008/2008-07-17-01.asp, accessed 6 February 2010

76. www.solarworld-usa.com/America-s-Largest-Sola.579.0.html, accessed 6 February 2010 and www.reuters.com/article/environmentNews/idUSTRE4AP50M20081127?sp=true, accessed 6 February 2010

77. Garthwaite, Josie (2009) 'Clean Energy Island: Maldives Goes Carbon Neutral, 16 March 2009, www.nytimes.com/external/gigaom/2009/03/16/16gigaom-clean-energy-island-maldives-goes-carbon-neutral-26038.html, accessed 6 February 2010

78. Walsh, Brian (2009) 'The Gusty Superpower. How Denmark's green energy initiatives power its economy', *Time*, 16 March

79. Negishi, Mayumi (2008) '"Solar City" proves allure of sun's energy in Japan', Reuters, 11 November, www.reuters.com/article/GCA-GreenBusiness/idUSTRE4AA2L620081111, accessed 6 February 2010

80. www.masdaruae.com/en/home/index.aspx, accessed 6 February 2010

81. Hinrichsen, Don (2008) 'Iceland Exclusive: Test Driving the World's Energy Future', www.peopleandplanet.net/doc.php?id=3233, 15 March 2008, accessed 6 February 2010

82. 'White House Goes Solar' (2003), *Solar Today*, March/April, p50

83. Mick Womersley (2004) 'White House Solar Panels at Unity College, Maine', www.moralequivalent.info/?p=29, accessed 6 February 2010

84. Pope, Carl and Rauber, Paul (2004) *Strategic Ignorance: Why the Bush Administration is Recklessly Destroying a Century of Environmental Progress*, Sierra Club, San Francisco, CA

85. Editorial (2006) 'Avoiding Calamity on the Cheap', *New York Times*, 3 November

86. Gore, Al, interviewed at *Meet the Press*, NBC, 20 July 2008, www.msnbc.com/id/25761899. Also, www.repoweramerica.com

Chapter 6 An Act of Dialogue

1. Tolstoy, Leo, quoted in Kanter, Rosabeth Moss (1977) *Men and Women of the Corporation*, Basic Books, New York, NY, p245

2. Morgan, George W. (1970) *The Human Predicament*, Dell, New York, NY, pxiii

Additional Resources

In the Notes I have listed books and other resources relevant to the particular topic being discussed. Here are some additional sources for information and education, both at the grassroots and international levels. A few are repeated for convenience. The selection also provides links to a wide variety of other resources worldwide.

International Solar Energy Society

Wiesentalstr. 50, D-79115, Freiburg, Germany

www.renewableenergyfocus.com

ISES is a membership organization with national sections in more than 50 countries; publishes *Renewable Energy Focus*, an online magazine on all aspects of renewable energy and activities around the world, with free subscription.

American Solar Energy Society

2400 Central Avenue, Boulder, Colorado 80301, USA

www.ases.org/solar

ASES is a membership organization with chapters or affiliates throughout the country; publishes *Solar Today*, a magazine with articles on all aspects of renewable energy, with reader services and product information; also publishes newsletters, books, booklets, conference proceedings and reports; sponsors conferences, forums and solar events such as the annual solar conference and the National Tour of Solar Homes.

Northeast Sustainable Energy Association

50 Miles Street, Greenfield, MA 01301, USA

www.nesea.org

NESEA is a membership organization; publishes *Northeast Sun*, a magazine; sponsors conferences, workshops, forums and renewable energy exhibits and events such as the annual Building Energy Conference, Solar Home Tours and the American Tour de Sol, an annual solar car race; maintains Greenfield Energy Park, a 5000m^2 greenspace, owned by the Town of Greenfield and built in partnership with NESEA, which holds concerts, workshops and demonstration exhibits of renewable energy items.

Union of Concerned Scientists

2 Brattle Street, Cambridge, MA 02138-9150, USA

www.ucsusa.org

UCS is a national nonprofit organization, with branch offices in Washington, DC and Berkeley, CA, dedicated to advancing responsible public policies in areas where technology plays a critical role; publishes *Nucleus*, a magazine, and many books, reports, brochures, briefing papers and visual materials, many of which are energy related.

Interstate Renewable Energy Council

P.O. Box 1156, Latham, NY 12110-1156, USA

www.irecusa.org

An educational and advocacy organization with a wealth of information and educational materials; conceived and promotes such programmes as Homes Going Solar, Communities Going Solar, Schools Going Solar and Businesses Going Solar.

Midwest Renewable Energy Association

7558 Deer Road, Custer, WI 54423, USA

www.the-mrea.org

MREA is a grassroots, nonprofit organization; activities include an annual Energy Fair; *ReNews*, a quarterly newsletter; workshops, programmes and policy development; events like the Fall Tour of Alternative Energy Homes.

Florida Solar Energy Center

1679 Clearlake Road, Cocoa, FL 32922-5703, USA

www.fsec.ucf.edu

FSEC's Renewable Energy Training and Education Center offers courses and workshops, both for local and international participants; its Document Sales Office offers more than 50 publications, slide libraries, videotapes, house plants and software.

Solar Living Institute

13771 South Highway 101, Hopland, CA 95449, USA

www.solarliving.org

Established in 1998, the Solar Living Institute is a nonprofit educational organization whose mission is to promote sustainable living through inspirational environmental education; it is a spin-off from Real Goods Trading Company and headquartered at its Solar Living Center, a 48,600m² renewable energy and sustainable living demonstration site; the Institute provides practical, education by example and hands-on workshops on renewable energy, green building, sustainable living, permaculture, organic gardening and alternative, environmental construction methods; it also offers internship opportunities to applicants from around the world.

Solar Energy Industries Association

805 15th Street Northwest, Washington, DC, 20001-2109, USA

www.seia.org

SEIA represents the solar energy industry in the United States; publishes *Solar Industries Journal*, the most up-to-date source for news about the photovoltaic and solar thermal industries, on both industrial and political fronts.

Solar Energy International

76 South 2nd Street, Carbondale, CO 81623-0715, USA

www.solarenergy.org

A pioneer in renewable energy education and sustainable development, SEI offers courses and hands-on workshops and training on all aspects of renewable energy, worldwide.

Grupo Fenix

Programa Fuentes Alternas de Energía, Universidad Nacional de Ingeniería (UNI), Apartado Postal LM139, Managua, Nicaragua

www.grupofenix.org

A nonprofit organization with year-round programmes and activities, Grupo Fenix offers the Solar Culture Course, an 11-day, twice a year programme of hands-on courses, workshops and service trips for individuals interested in broadening their knowledge of renewable energy and sustainable development; participants from around the world live in a rural village in Nicaragua while working side-by-side local community members as they teach them about practical solar energy technologies such as solar cookers, solar dryers and photovoltaic systems; technical classes are taught by Dr Richard Komp, an expert from the US with over 40 years experience working with solar technologies, who is President of Maine Solar Energy Association and author of *Practical Photovoltaics: Electricity from Solar Cells*.

Appropriate Infrastructure Development Group

33 Harrison Avenue, Boston, MA 02111, USA

www.aidg.org

With projects in Guatemala and Haiti, AIDG helps individuals and communities get affordable and environmentally sound access to electricity, sanitation and clean water; through a combination of business incubation, education and outreach it helps people get technology that will better their health and improve their lives; it has implemented projects utilizing PV, wind-turbines (locally constructed), biogas and microhydro systems.

Boston Area Solar Energy Association

33 Cambridge Terrace, Cambridge MA 02140, USA

www.basea.org

Since 1982, BASEA has been an active part of the region's energy-use conscience; an educational and voluntary organization, BASEA sponsors monthly forums on a wide range of solar energy topics; writes policy statements to affect the political process, and co-sponsors annual events such as Junior Solar Sprint (model solar cars designed, built and raced by elementary and junior high school students) and Boston Solar Day (an energy fair of renewable energy related activities and demonstrations).

Sunnyside Solar, Inc.

1014 Green River Road, Guilford, VT 05301, USA

www.sunnysidesolar.com

Sunnyside Solar, Inc. is a small family owned company specializing in photovoltaic electric systems – Sunlight to Electricity; founded in 1979, it has primarily been working in the design and installation of small to medium photovoltaic systems, both stand-alone and grid-connected; for the past several years, the owners, Richard and Carol Gottlieb, have been involved in education, originally giving workshops from their shop and with Solar Energy International, with more recent programmes with Ulster (NY) BOCES and Greenfield (MA) Community College; Richard, with a master's degree in Solar Energy and a background in physics and engineering mechanics, helped develop the separation device used to launch the instrument-bearing Vanguard satellite while being employed in the satellite structures group at the Naval Research Laboratories in Washington, DC, in 1959.

Massachusetts Energy Consumers Alliance

670 Centre Street, Boston, MA 02130-2511, USA

www.massenergy.com

A fuel co-op, founded in 1982, Mass Energy has pursued a mission to make energy more affordable and environmentally sustainable for the consumers of Massachusetts; it is also the state's premier organization in educating the public about, and for marketing of, renewable energy products; it offers several green power programmes, which enable consumers to choose electricity resources that are cleaner and healthier for the environment than current sources of their electricity

Solar Cookers International

1919 21st Street, Sacramento, CA 95811, USA

www.solarcookers.org

A nonprofit organization committed to spreading solar cooking awareness and skill worldwide; publishes *Solar Cooking Archive*, an internationally recognized internet resource for solar cooking information; also publishes a catalogue of solar cookers for sale, plans and instructions for building solar cookers, and solar cookbooks.

Solar Design Associates

P.O. Box 242, Harvard, MA 01451-0242, USA

www.solardesign.com

The company is a world leader in photovoltaic utility-interactive system design and installation; its president, Steven Strong, designed, among many other projects, the world's first solar electric neighbourhood in the town of Gardner in Massachusetts; he is also the author of *The Solar Electric House.*

Global Transition Group

55 Middlesex Street, North Chelmsford, MA 01863, USA

www.globaltransition.net

The GT Group consists of the following organizations committed to the global transition to a sustainable energy future: Enersol Associates, Inc., an internationally recognized NGO leader in the development and introduction of sustainable energy innovations for rural development; Soluz, Inc., the developer of one of the world's leading PV commercialization projects, providing solar-generated electricity for a monthly fee to rural customers in developing countries; and Global Transition Consulting, Inc., a consulting firm providing project and business development advisory services to foster the global transition to sustainable energy systems; the Group grows out of the work of Richard Hansen, who has been a pioneer in solar technology applications.

Solar Electric Light Fund

1612 K Street Northwest, Washington, DC 20006, USA

www.self.org

SELF, a nonprofit corporation, plays a major role in introducing solar electric systems around the world; it has also provided the basis for the formation of such companies as SELCO (Solar Electric Light Company, Washington, DC), SELCO-India, SELCO-Vietnam and joint ventures in solar distribution and manufacturing in China and Sri Lanka; it also has completed projects in South Africa, Indonesia, Nepal, Uganda, Tanzania, Solomon Islands and Brazil.

350.org

www.350.org

Founded by Bill McKibben, environmentalist, activist and author of *The End of Nature*, 350.org is an international campaign that's building a movement to unite the world around solutions to the climate crisis. In October of 2009 it coordinated 5200 simultaneous rallies and demonstrations in 181 countries. It is organizing a 'global work party' all over the world to take place on 10/10/10 to put up solar panels, dig community gardens, and other solutions, calling upon world leaders to do the same.

Index